The Gyroplane Flight Manual

For Gyrocopters and Sport Gyroplanes

by
Paul Bergen Abbott

All photographs and illustrations are by the author unless otherwise identified.

Published by
The Abbott Company, Indianapolis, Indiana USA

ISBN 1-888723-00-9

Dan—

May you always fly in sunshine.

Contents

What Are You

Your life will never be the same.

When you learn to fly a gyroplane, you accomplish something important. You aren't simply learning a new skill, like riding a bicycle or playing a saxophone. No, indeed. You're becoming a different person.

Whether you start as a licensed pilot or with absolutely no knowledge of flying, you're sharing the pioneering spirit that only a few people knew in the past. You've heard the names of those people: the Wright brothers, Glenn Curtiss, Otto Lilienthal and the inventor of the gyroplane, Juan de la Cierva. You've heard about the fabulous things those people accomplished. What they did is what you're about to do, jumping into an experimental flying machine and trusting it to lift you into the air.

If you were the first person to do what you're doing, *your* name would be the one in the history books! This isn't an off-the-shelf ready-made one-size-fits-all experience. You don't have to share the credit with someone else. This is your show. You're the star!

Getting Into?

You have all the help you need. Your instructor sits with you at first. A designer gives you the plans. Friends offer advice and support. But in that first solo flight, no one but you holds the stick. In later flights, nothing but your efforts keep your gyroplane airworthy. *You* do the flying. *You* know the feeling. And *you* are the one who thinks, "Yes, Orville and Wilbur, I know what it's like. I'm sharing these experiences with you!"

Chapter 2

Meet the Gyroplane

Paul Bergen Abbott,
who wrote this book,
flies his Bensen B8-M Gyrocopter

A gyroplane is more than just an aircraft. It's a thrilling, exhilarating experience. It's a new perspective on the world. It's fun!

If you ask a gyroplane pilot what's most significant about this machine, he'll talk about the experience of flying it. He'll tell you how this assembly of ordinary parts gives him an elevated perspective on the world, changing his viewpoint on life and making everything in the world just a little more beautiful from the air. It makes him feel successful, and gives him a feeling of pride that's balanced by a humbling familiarity with the powerful forces of nature.

A gyroplane is made of what would otherwise be a pile of ordinary materials: some aluminum, some steel, some bolts, some wheels and maybe even some plastic. But when these parts become a gyroplane they're given almost magical qualities. They become able to lift a human being, raising him away from the ground on a cushion of invisible air, giving

6

PHOTO BY DANIEL ABBOTT

him the ability to move like Superman, flying wherever and however he wishes, at speeds too slow or fast for any earth-bound human to duplicate.

What's good about gyroplanes?

What is it about gyroplanes that fascinates people? Among sport aircraft, there's one main appeal: This is a unique flying experience. Being airborne on spinning rotor blades is different. Since those blades are always moving, the gyroplane can stay in the air at any speed. Stalls, as in an airplane, are unknown among gyroplanes.

Gyroplanes are very maneuverable. They can turn tightly, close to the ground, to stay over a small area. Steep banks that are considered aerobatic in an airplane are normal play for many gyroplanes. Punch the pedals and a gyroplane will turn to fly sideways (the rotor doesn't care which direction the body points). Pour on the power and you'll usually find an impressive rate of climb. This is a type of aircraft that loves to play.

Jerrie and Erlene Barnett enjoy a flight in their Barnett J4B2

Homebuilt gyroplanes are typically folding-wing aircraft. Their rotor blades attach to the craft with a few bolts that you can easily remove. That means you can haul a gyroplane on a trailer and store it in a garage.

Some—not all—gyroplanes are inexpensive, both to buy and to maintain. Their cost is concentrated in the rotor and control systems, so the rest of the machine can be simple in design and materials. A bolt-together airframe is good enough for several of the gyroplane designs, and a stream-lined body pod can be easily mounted over the frame to complete the picture.

Gyroplanes can fly in strong winds that keep airplanes on the ground. A good pilot can take advantage of the gyroplane's ability to ignore the effects of wind and fly on days that are uncomfortably gusty for small airplanes. Why? Gyroplanes all have the equivalent of high wing loading, with their small-chord rotor blades doing the work of a much larger-chord fixed wing. It's more like riding a Bonanza than a Cessna 152. And gyros usually don't have much of a fuselage (some have no fuselage) to catch a crosswind. When a crosswind or gust hits, the rotors can keep generating lift regardless of the direction of the wind.

What's bad about gyroplanes?

Gyroplanes have some disadvantages. You won't find a homebuilt gyroplane winning an efficiency race. Rotors aren't efficient, since they produce more drag than an airplane wing. They won't go very fast, either. That's partly because it takes a lot of power to push a draggy rotor disc through the air. It's also an unavoidable result of having a rotor blade that's always spinning at 300 to 400 miles per hour. When you try to add a lot of forward speed to the aircraft, you're asking the advancing rotor blade to go even faster— a tall order!

It's possible to make gyroplanes fairly efficient, but for sport flying, efficiency isn't really a very important concern.

Is it true what I heard about them?

There are a lot of myths about gyroplanes. Like "They're dangerous." "They're unproven." "They have a lot of unknown flight characteristics." Not true.

Gyroplanes are no more dangerous than other homebuilt aircraft. It's the builders and the pilots who are the

The Farrington Twinstar

problem in both gyroplanes and airplanes. Anything that flies can be dangerous, regardless of whether it beats the air into submission with whirling rotor blades or carves glidepaths through the air with wings.

Gyroplanes seem to attract a large number of tinkerers without much understanding of aircraft mechanics or construction techniques, and some of them have done some foolish things. Also, when a gyroplane tips over, the spinning blades usually splinter themselves, making a spectacle of an otherwise uninteresting incident.

Handled right, homebuilt gyroplanes are at least as safe as any other homebuilt aircraft. Statistics show that, for a competent gyroplane pilot, the most dangerous part of a flight is the automobile trip to the airport!

How is a gyroplane different from a helicopter?

The difference is simple: helicopters drive their rotors with engine power, autogyros have free-wheeling rotors. But that simple difference changes the whole flying experience.

The helicopter is more complex mechanically, requiring more maintenance and demanding more pilot skill than other types of aircraft. But the helicopter does something no other aircraft can: it lifts off from your back yard, from an unprepared field, or anywhere you like. And it can move through the air virtually any way you like: forward, backward, sideways or even completely stopped in a hover.

The gyroplane (also called "gyrocopter," "autogyro" or just "gyro," depending on what it looks like and who's talking) is usually a very simple flying machine. There's no tail rotor, and one less control than a helicopter (no collective), so you can fly a gyro with stick, rudder and throttle like an airplane. The traditional open-frame gyrocopter is about the simplest thing in the air next to a boomerang. There's not much maintenance required (although the little maintenance that's needed is critical), and some models are fairly low in cost. The open-frame style gyro is one of the most minimal of flying machines, giving the pilot a very birdlike experience.

Can a gyro be flown as an ultralight?

Yes, you can fly a gyroplane as an ultralight if you like. There are a number of autogyro designs that can operate under the ultralight rules in FAA regulations Part 103: the Air Command 447, the Brock KB-3 and the Gyrobee, for ex-

The Gyrobee flies as an ultralight, using a 47-horsepower Rotax 447 engine.

ample. The FAA wrote their regulations for "ultralight aircraft," not "ultralight airplanes." Some gyros have been specifically designed to meet these rules and to allow the pilot to fly without a license. Ultralight gyros have been around since 1984, and by now some of them

10

can provide good-enough performance to win over some former ultralight airplane pilots.

To keep a gyro within the 254-pound ultralight weight limit, you can't load on all the accessories you might want. For example, a prerotator to pre-spin the rotors before takeoff isn't excluded from the weight limit by the FAA (Although many people think this is safety equipment that should be excluded like a ballistic parachute, the FAA has not yet agreed). That means you may have to push the rotor blades by hand to start them.

An ultralight gyro will usually have a small engine, like a Rotax 447, and could end up with anemic performance. The little power plant is usually needed both to keep the weight down and to keep the top speed under the ultralight maximum of 63 miles per hour. If the pilot is a big person, this may not be enough oomph for good performance. However, some more average-sized pilots are reporting satisfactory performance from a 447. And the Brock KB-3 gyro is light enough to use a big 65-horse-power Rotax 582, limiting the top speed by increasing the pitch of the rotor blades.

What other choices are there?

Do you prefer to fly with the wind in your face? Or completely surrounded in cabin class?

With gyroplanes you can have anything from a bugs-in-your-teeth open-frame traditional gyrocopter to a sealed-up snuggly-warm RAF 2000. In the middle there are also partially-enclosed machines like the Air

An open-frame Air Command offers a feeling of freedom.

11

Command, giving you the feel of an airborne convertible with the top down.

The open-frame gyros are a real thrill! You sit on a simple plastic seat, your control stick in your hand, your feet on the rudder pedals in front of you—and the world below you! The wind massages your face and ripples your pants legs, you hear the sound of your engine behind you and the swish of the rotor blades above you. This is as close to an addictive experience as you can legally get, and some devoted open-frame pilots wouldn't have it any other way.

If you go for a partially-enclosed gyroplane, you can either go topless or sideless. You can take the topless approach in an Air Command, with a body pod that's like an open-cockpit airplane. Or you can take the sideless approach in a Sport Copter Vortex with no doors.

Gyroplanes give you another choice: which engine should you pick? Homebuilt gyroplanes use many of the same engines that are found on airplanes. There are two-strokers like the Rotax, Arrow and Hirth, and four-strokers like the Volkswagen and Subaru. If you insist on a type-certificated aircraft engine, you can find gyroplanes that use

The Sport Copter Vortex has a partial enclosure.

George Charlet and a young friend enjoy the thrill of a gyroplane flight.

a Continental or Lycoming. Some gyroplanes use engines you don't find anywhere else, like the two-stroke McCulloch drone engine that gyrocopter pilots love. Gyroplane people have always been adventurous in trying new engines, due to the excellent engine-out landing capabilities of these machines. They can glide to a landing in a very small area, touching down at about zero airspeed.

How do you get started in gyroplanes?

Today it's as easy to get started in homebuilt gyroplanes as it is in homebuilt airplanes. Flight training in homebuilt gyroplanes has become widely available in the past few years. Since the late 1980's, powered two-seat training gyroplanes have appeared, along with certified flight instructors for gyros. Today you can hire an instructor to teach you to fly a gyro with all the safety of dual-seat instruction. Previously, gyro pilots had to take a very risky "teach yourself" approach that led to a high accident rate among new pilots and a bad rap for the gyro crowd. But now it's different, and the availability of gyro training has generated an explosion of interest in these machines.

To learn to fly a homebuilt gyroplane, you must have training. This is true regardless of your previous flight experience. Fixed-wingers have to learn how to handle an aircraft that doesn't fly like an airplane. New pilots have to learn a whole new ball game.

Some gyroplane manufacturers offer dual-seat training. Pilots of other models can learn in a gyroplane that resembles the machine they intend to fly, then transition to their own craft. Several instructors teach in one of the open-frame two-seaters that resembles a Bensen gyrocopter.

What can you do with a homebuilt gyroplane?

To date, most homebuilt gyroplanes have been used in one way: as a sport machine to fly in a local area, often with other people with the same interest. But they're capable of much more. Homebuilt autogyros have been flown coast-to coast in the United States (by Ken Brock in an open-frame gyrocopter in 1971 and by Howard Merkel in a Barnett J4B-2 in 1989). Some pilots fly their homebuilt gyros regularly cross country, sometimes from state to state.

Homebuilt gyroplanes make up a rapidly-expanding part of the sport aviation scene. Every year there are more homebuilt autogyros and more people wanting to build and fly them.

Chapter 3

Picking the Best Gyroplane

Let's pick out the best sport gyroplane just for you!

I'm going to give you some advice that will simplify your task of checking out all the gyros that are available to you. After all, I know just what you want: the perfect gyroplane! Somewhere out there is just the right design, with just the right features and just the right performance to exactly match your preferences.

If you're like most of us, you can spend a lot of time looking at all these machines. You'll check out the specs on the ones that interest you and talk to people who have flown your favorites.

But chances are you'll find more than one gyro that turns you on. That makes it harder to choose one when there are two or three that seem like the answer to your dreams. If you're like me, you may want them all!

The trick to finding the very best gyro for you is to make the right choices. There are just a few things you have to decide, and if you can make each of those decisions right, you'll be left with the one gyroplane that's best for you (or several gyros that are all good for you, and you can "eeny meeny" and you'll be right, no matter what).

To help you, I'm going to list the choices and give you an example of a machine that fits each side of each choice. When you match them to your own tastes and preferences, you'll end up with a neat description of your best machine.

The choices

To choose the best gyroplane for you, here are the basic choices you need to make:

- Open-frame or enclosed cockpit?

- Single-seat or two-place?

- Gyroplane or helicopter?

- Ultralight or licensed?

- High-profile design or low-profile?

- Established design or new?

Open-frame or enclosed cockpit?

The first choice I'll tackle is the one between bugs-in-your-teeth flying vs. cabin class flying. For this comparison, I'll use two rotorcraft that I've flown in: the **Brock KB-2** and the **RAF 2000**. Actually, I haven't flown a real Brock KB-2, but I've flown a machine very much like it, a McCulloch-powered Bensen Gyrocopter. Although I haven't soloed an RAF 2000, I've taken the stick while flying with its developer, Dan Haseloh. And, as you'll see, that's enough experience to give you a comparison.

Open-frame: **Brock KB-2** is as open as an aircraft can be. You sit on a seat that's not enclosed in any way. There's nothing on the left of you, nothing on the right, and

you see only your feet and a small framework in front of you. The engine and aircraft are above and behind you where you can't see them while flying. In the air, you have the feeling that you're in some sort of magical flying seat that moves at your command. The wind in your face is strong, but on warm days it's cool and refreshing. The tugs and pulls of the air on your body as you move along give you a sensation of flight that you can't get in an enclosed machine. You must wear a helmet in an open machine like this, but you soon discover that the little window of your helmet opens onto a scene of grand scope.

Enclosed-cockpit:

RAF 2000 is enclosed in a plastic bubble that shields you from the wind. You can fly with the doors on, giving you complete separation from the inconvenience of the breezes. And you can turn on the cabin heat and keep flying right through the winter, as they do in Saskatchewan, Canada where the RAFs are made. You can take the doors off during the heat of summer, giving you a semi-enclosed feel and more sound of the wind, the rotor blades and the engine. The transparent plastic bubble around you gives you a feeling of security without seeming to close you in. RAF pilots enjoy a level of flying comfort normally found only in factory-built airplanes.

The choice

Open-frame: Pick the open-frame machine if you prefer riding a motorcycle to driving a sports car—or if you want to feel a unique thrill of flight that can't be matched by any other aircraft. The open-frame rotorcraft is for the adventurer, trading comfort for contact with the elements. Without an enclosure, these open machines are also generally lighter and more responsive to your control. It's almost like flying without an aircraft.

Enclosed-cockpit: Pick the enclosed machine if you like comfort. That may be especially important if you fly in cold weather or if you want to spend long periods of time in the air, as on cross country flights. Enclosed machines typically allow you to start up everything from the pilot's seat. They must have prerotators for the rotor blades and electric starters for the engines, since you can't reach them from inside the cockpit. The enclosed machines also have an extra measure of crash protection, since the enclosure provides a protective cocoon around the pilot. Many pilots fly enclosed machines without helmets. While this is a safety compromise, many people feel it's an advantage of the enclosed gyroplane.

Single-seat or two-place?

For this comparison, let's use the **Air Command 582** and the **Parsons Trainer**, two of the oldest and most popular examples of each type.

Air Command 582
leaves you
totally alone
in the sky. No
one interferes
with your
flying (there's
no back-seat
driving from
your mother-
in-law), and
you experi-
ence a feeling
of oneness
with your

machine, since it's just you and your rotorcraft. You do what you want in the air without ever having to think about someone else in the machine with you. The single-seat Air Command is light and responsive. With only your weight to carry, the machine climbs easily, maneuvers rapidly and sips very little fuel.

Two-place:

Parsons Trainer

carries you and a friend. You have someone to talk to while you're flying, and with an intercom it's fun to share the experience. Its heavier weight makes it feel smoother and more stable.

The feel of the controls is slower and gives you more time to react to bumps and changes in the machine's attitude. Two-seaters were originally designed for instruction, and most two-seat flying is with an instructor.

The choice

Single-seat: Pick the single-seat machine if you want to play. If you want responsiveness, quickness and a strong feeling of control, go solo. If you want to fly as an ultralight, you must fly a single-seater, according to the rules. Ultralights aren't allowed to carry passengers except under special training situations. Even if you have a student pilot license, you'll still have to fly without a passenger. If you have to fly alone, it's better to be in a smaller, lighter and generally cheaper single-seat machine. Learning to fly is more difficult and risky in a single-seater, since you can't take dual instruction in it. You learn to fly in a bigger dual-seater and then must climb out of your trainer and go it alone. That's the most difficult and risky part of learning to fly a homebuilt rotorcraft.

Two-place: Pick the two-seat machine if you have at least a private pilot license with a rotorcraft-gyroplane rating. You must have at least that license in order to carry someone else in the second seat. Also, make sure you're a really good, and experienced, pilot and mechanic. You're taking responsibility (and legal liability) for your passenger, who is trusting you totally. You can fly solo in the same two-seater in which you take instruction, and the regulations allow you to hire a CFI to train you, so long as you own the trainer.

Gyroplane or helicopter?

I really shouldn't touch this hot potato! The feelings run strong on both sides of the debate over whether or not you should power your rotor. But if you're considering a gyroplane, you're probably also wondering about a helicopter. Today you really do have this choice, since several homebuilt helicopters are available and more are being developed all the time. Representing the gyroplane team this time is the **Barnett J4B2**. And on the helicopter side it's the **RotorWay Exec**.

Gyroplane:

Barnett J4B2 is the latest version of one of the oldest sport gyroplanes around. While Igor Bensen was developing his two-stroke-engine-powered Gyrocopters, Jerrie Barnett

was mounting a big old Continental certificated aircraft engine on a bigger airframe. The Barnett machine is fully capable of comfortable cross country flight (Howard Merkel flew one from coast to coast in 1989). This gyroplane is relatively easy to fly; most people would say it's easier than an airplane. The maintenance is simple, since the rotor head and blades are of a simple design and are not under high stress. The prerotator allows you to take off from a much smaller space than an airplane (the takeoff distance varies greatly depending on the wind and the density altitude) . But you do need some sort of a runway, or at least a grass strip of at least, say, 800 feet for a gyro.

Helicopter:

RotorWay Exec was introduced in the early 1980's. It was at least the third-generation product of B. J. Schramm, who first flew a Bensen Gyrocopter and then wanted the

PHOTO COURTESY ROTORWAY INTERNATIONAL

advantages of a true helicopter. This helicopter will fly cross-country on its RotorWay engine that's specially made for it, as was proven when Homer Bell flew his Exec from coast to coast in two separate trips. As a true helicopter, this ship is fully capable of taking off from your back yard (if you have tolerant neighbors). Compared to a gyroplane, any helicopter, including the Exec, is more difficult to fly. Exec pilots spend many, many hours practicing each element of flight before moving on to the next step. They also religiously practice engine-out autorotations, which are considerably more critical than in an gyroplane. Maintenance of the more complex Exec is more demanding than the simpler gyro.

21

The choice

Gyroplane: Pick the gyroplane if you want a simpler machine to build and maintain, and if you don't mind operating from airports or landing strips. The gyro is generally cheaper to buy, maintain and operate than a helicopter. But with a gyro you'll spend the rest of your flying days answering the question, "What kind of a flying machine is that?" Nobody but an aviation nut knows what a gyroplane is, and even fewer have a good idea of how it works. To a lot of us, being unusual is an advantage!

Helicopter: Pick the helicopter if you want to do things no other aircraft can do, like taking off from an unimproved field or a small spot. Despite the best attempts of gyroplane people to fake it, the helicopter is the only homebuilt aircraft that can hover. It's the only aircraft that can fly into and out of a really small space. The helicopter is harder to fly and it's more critical and difficult to maintain, but if you're a person who can handle that, the helicopter will reward you with flight above and beyond the capabilities of any other aircraft.

Ultralight or Licensed?

This choice isn't simply a matter of whether or not you already have a pilot's license. It also depends on what you want to do with your gyroplane. To find out, let's compare the ultralight **Brock KB-3** and the licensed **SnoBird 582**.

Ultralight:

Brock KB-3 was designed especially for the person who wants to fly a gyro as an ultralight. It gets its light weight of only 250 pounds by leaving off an enclosure,

a common practice among ultralight gyros. It also uses relatively lightweight aluminum rotor blades. Unlike most other ultralight gyros, the KB-3 has a big engine (a 65 horse-power Rotax 582). To keep this big power plant from pushing it past the maximum ultralight speed of 63 miles per hour, the lightweight aluminum rotor blades are set at a high pitch. This, in turn, requires a prerotator to pre-spin the blades, so the KB-3 has a lightweight belt-driven unit. The big engine and prerotator are very helpful, especially to a new gyro pilot, but are typically left off ultralights to make the weight limit of 254 pounds. Like all ultralights, the KB-3 fuel tank is limited to five gallons.

When this gyro is operated as an ultralight, the pilot doesn't need to have a license. This means no medical exam is required and no paperwork is needed. An ultralight gyro won't have an FAA inspection and won't be restricted to a test area for your first flights. You don't have to register it any-where, which probably also means you won't pay the aircraft tax that's imposed in many states. You can legally buy an ultralight gyroplane already built (from an individual; the Brock factory only sells kits) without having to fabricate 51 percent of it, as is required of licensed machines. And you yourself can maintain that ultralight you just bought.

Licensed:
SnoBird 582

has more power and more range than the ultralight. It can carry as much fuel as you like since it's not limited to the ultralight maximum capacity of five gallons. It

PHOTO BY DANNY WHITTEN

can fly fast, not being restricted to the 63 mph top speed of ultralights. As a licensed aircraft, the SnoBird carries a strong prerotator that pre-spins the blades very effectively. Without a weight limit, you can also install brakes, mufflers and many other accessories. You yourself must have a license to fly this licensed machine, and you can fly it into any airport that receives support from public funds (most of them do). As a licensed aircraft, you have the reassurance that this machine has passed an FAA inspection that usually requires standard aircraft construction techniques and is considered an airworthy machine. The FAA inspection doesn't guarantee safety, but it's more than the no-inspection situation of ultralights.

The choice

Ultralight: Pick the ultralight if you don't want to bother with the FAA or with paperwork. This may be the option for you if you never intend to carry a passenger (passengers are illegal for ultralights). Remember that as an ultralight pilot, you're still responsible to the FAA. They can inspect your ultralight aircraft at any time and still require you to observe the "rules of the road" regulations of Part 91. Ultralight pilots need as much training as licensed student pilots, but the training doesn't have to be from an FAA-licensed instructor. The ultralight association and Experimental Aircraft Association have their own instructor certification programs.

Licensed: Pick the licensed gyroplane if you already have a license or intend to get one; if you're a large person who needs more power to fly; if you want to carry a prerotator, brakes or other accessories; or if you want to carry enough fuel for long cross-county flights or carry a passenger. If you want to fly from an airport where there's other traffic, you'll probably find that the licensed machine is more accepted. The licensed gyroplane pilot can add all his/her rotorcraft flight time to the logbook, building up hours at a very favorable cost.

High-profile or low-profile?

This is a relatively new category, which began with the first design that could clearly be described as high-profile, the **Dominator**. To give you an idea of the purpose of this new-type design, let's compare it to a low-profile gyro like, say, the **Parsons single-place**.

High profile:

Dominator strikes some people as an unusual-looking machine, with its long landing gear giving it a mosquito-like appearance. Because it stands up high, this machine can

swing a big-diameter prop to get more thrust out of the Rotax 503 engine (or 582 if you prefer). Before takeoff, the pilot sits much higher above the ground (in the air, naturally, there's no difference) and the pilot's position is intended to put his weight right on the center of thrust. This high-profile design is intended to provide stability, efficiency and good pilot handling qualities. It has a fully-articulated "tall tail" which is intended to minimize the propeller's "P-factor" or "torque" (the tendency to make the nose want to go right or left at high power).The machine is also said to be highly resistant to pilot induced oscillation (PIO or porpoising) and resistant to a disastrous characteristic, the power pushover.

25

Low-profile:

Parsons single-place

gyro sits about as low to the ground as a gyro can get. The wheels are mounted on axle tubes that are bolted right onto the bottom of the keel. The pilot sits in an upright position, somewhat lower than the center of thrust, and can easily jump in and out of the seat if needed to hand start the engine or to pat up the rotor blades. This gyro would be regarded by most people as conventional-looking (except for those Dominator fans who regard the tall look as normal). The tail is a simple fin/rudder unit (often a Brock KB-2 tail) which is mounted simply on the end of the keel.

The choice

High-profile: Pick the high-profile design if you want something a little different. In this machine, you'll look different from almost anything else that can be found in the sky. You'll have many questions to answer from onlookers when you show up to fly. You'll have a stable machine with docile handling qualities and good power from your Rotax 503. If you ask somebody who flies one of these high-profile machines, it's unquestionably superior to the low-profile designs! (But read on...)

Low-profile: Pick the low-profile design for the configuration that's the standard of the homebuilt gyroplane world. You'll already have enough questions to answer about what sort of a flying machine this is; with the low-profile you can at least avoid some of them. Low-profile designs come with various handling qualities. If you ask somebody who flies a low-profile machine, it's unquestionably superior to the high-profile design!

Established *vs.* new

Luckily, this won't be a choice for most of us—whether to pick an established design with many copies flying for several years or a brand new design that's just coming out.

For most of us, the established-vs.-new choice isn't a concern. Most rotorcraft people will go for a design that's at least reasonably well established. For some of us, that means one that has been out for many years and has survived the tough test of being built by many different people, probably crashed a few times, and with improvements already made.

For others it's enough just to see a prototype flying well, to look it over and to decide that its design and construction are good. That's exciting, and that's in the spirit of homebuilt experimental aircraft. But someone who takes on a new design should recognize that there may be delays in its development and corrections that need to be made in the design after multiple copies of it are flying. When you're one of the first to adopt a new design, you become, in effect, a co-developer of the new aircraft.

As a general rule, I suggest that only more experienced people should take on the brand new gyroplane, while new pilots will be better off picking a more established design.

Abbott's recommendations

When you think about all these comparisons, there are some recommendations that I can make to everyone:

First, it's important that you consider your own knowledge and skill. If you don't know much about aircraft construction or if you're all thumbs, I believe you're better off sticking with proven rotorcraft from manufacturers who offer extensive builder support.

Second, consider the knowledge and skill of your friends. They can help you, telling you what you don't know and helping you with the more difficult parts of construction. Joining a local chapter of the Popular Rotorcraft Association is a great way to boost this "friend factor."

Finally, be realistic about what sort of flying you'll do. You may fantasize about flying out of your back yard, but very, very few people ever actually do it. You may imagine commuting to the roof of your business, but almost no one has ever done that. You may plan on taking up your best girl or your wife and kids on lazy Saturday afternoons, but that's a big, big responsibility and only a few people actually do that. Most gyroplane pilots fly simply for the sport of it, operating from places where other like-minded people are also flying homebuilt rotorcraft.

I'll be a monkey's uncle if you can't find just the right gyroplane for you!

Chapter 4

Where Did the Gyroplane Come From?

A gyroplane is unique and different from other aircraft, but it's not strange and it's not very new. If a gyroplane looks odd to you, it's your viewpoint that is odd, not the nature of this machine. A gyroplane is a highly developed aircraft, with all the components working properly together, just like an airplane or an automobile or a lawn mower. The lawn mower that's so familiar to you would seem like some kind of wild, violent contraption to a primitive tribe in New Guinea. To those natives it's indeed a strange device, but to you it's common enough to be pretty boring sometimes. Likewise, a gyroplane may not be familiar to you, but it's not strange, having a long history that goes back at least to the 1950's.

The origins of sport gyroplanes

The recognized developer of the amateur-built sport gyroplane is Igor Bensen, a Russian-born engineer who coined the name Gyrocopter for a rotary-wing aircraft he designed. The name Gyrocopter was originally a trademark until Bensen Aircraft Corporation, the manufacturer of Gyrocopters for over thirty years, closed operations in 1988. Now that term is sometimes used

to refer not just to Bensen-design machines but also to other similar aircraft, now available from several manufacturers.

The word "gyroplane" is the name used by the Federal Aviation Administration to describe the whole category of aircraft with rotor blades which are not directly powered by the engine. (Curiously, the word "gyroplane" came from the name of the first helicopter to fly with a person aboard, a machine made by Louis Charles Breguet that flew in France in 1907. Its rotors were driven by an engine, so today it would not be classified by the FAA as a gyroplane.)

You may also hear the word "autogyro" around these aircraft. That was the name of the first successful rotorcraft (originally spelled "autogiro" as a trademark). It was the direct ancestor of today's gyroplanes and modern helicopters. An autogyro had an airplane-like fuselage with the engine in front. Today many people call any rotorcraft with an engine on its nose an autogyro.

Igor Bensen was a pilot and engineer on those early autogyros. He was one of the small number of flyers who had experience with these machines, and developed enough expertise to be asked to head the evaluation of one of the special rotorcraft that were developed during World War II.

To help with aerial observation from German submarines during the war, the Focke-Achgelis company built a small rotorcraft that could be towed by a rope from a submarine. Seated several hundred feet above the U-boat, the pilot of the rotorcraft had a bird's-eye view of things for miles around. The aircraft had a small three-bladed rotor above it that turned in the wind in autorotation. The pilot sat on a simple open frame

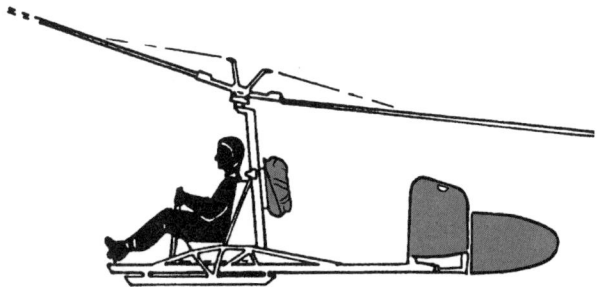

Fa 330
German gyro towed by a submarine

and steered the machine with a control stick. When he spotted trouble he reported it on his telephone. Then the crew winched him in to land on the sub. He then folded the rotors and the rest of the machine into a small bundle and climbed aboard in time to submerge.

The British had their own version of this type of aircraft. The Rotachute had a two-bladed rotor, a very simple airframe and no engine. It was this towline rotorcraft that the General Electric Company was asked to evaluate in 1946, with the young Igor Bensen in charge.

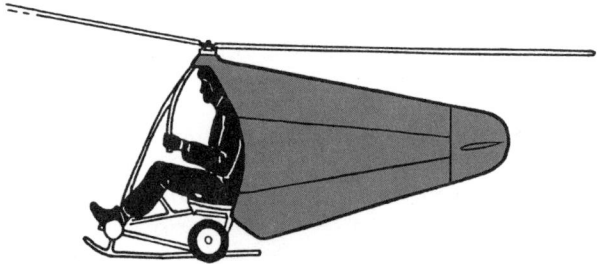

Rotachute
British towed glider

Apparently Bensen was impressed with the idea of the Rotachute, because later, in 1953, he formed his own company and began to work on an engine-powered version of that concept. Bensen mounted an engine in a pusher position behind the pilot and, after several successive designs, finally produced the machine he called a Gyrocopter. The first aircraft to bear this name was his model B-7, introduced in 1955. It had an airframe made of round aluminum tubing, wooden rotor blades and a modestly-powered 42-horsepower Nelson two-stroke engine. It was controlled by an overhead stick hanging down from the rotor head. The rotor head itself was a simple spindle projecting up from a bearing. It required good pilot proficiency to fly this machine well.

It wasn't an easy task to develop the Gyrocopter. Bensen and his co-workers had to do their own designing, trouble-shooting and test flying. Bensen had no one to teach him to fly his machine, since no one had ever built this type of aircraft before.

Also, he was working in the post-war decade of the 1950's, when the autogyro was considered to have been outmoded by the newly-invented helicopter. And Bensen's simple, open-frame rotorcraft was almost a complete opposite of the larger, faster post-war airplanes that were drawing public attention.

Bensen the designer turned into Bensen the public relations promoter and generated publicity in magazines and newspapers (This was before there was TV!). He got the support of the Reynolds aluminum company, which helped promote this machine made of their materials.

Igor Bensen built variations of his machine with boat hulls and floats to fly in the water shows at Cypress Gardens. He flew public demonstrations under challenging conditions of weather and air traffic, narrowly avoiding disaster in several cases, such as the time he was nearly run over by a military jet precision flight team.

Igor Bensen persisted with his Gyrocopter, and introduced a refined model in 1957, the B-8M (M stood for motorized. There was also an engineless B-8 gyroglider for training purposes). It had a fuselage made of two-inch-square aluminum tubing and it carried the bigger McCulloch engine. With 100 cubic inches of displacement, this two-stroke power plant was rated at 72 horsepower, enough to finally make the Gyrocopter a performer.

Dr. Igor B. Bensen, inventor, designer and test pilot, flies his B8-M Gyrocopter.

That 1957 model was the basic aircraft produced by the Bensen Aircraft Corporation for three decades. Over the years it had many important refinements, including the offset gimbal rotor head, metal rotor blades, an optional joystick control, a 90 horsepower version of the McCulloch engine and others. Bensen also built and flew prototypes of several other models, including a larger two-seat machine, but these were not put into production.

Along with inventing the Gyrocopter, Bensen started the Popular Rotorcraft Association, which exists today as the international organization for sport rotorcraft enthusiasts. He developed training methods for learning to fly Gyrocopters and lobbied to create a special regulation for Gyrocopters in FAR Part 101, which used to allow a person to be legally licensed to fly solo on a Student Pilot license simply by demonstrating three takeoffs and landings to an instructor in a towed gyroglider (That rule is no longer in effect. Today, to fly a licensed gyro you must be signed off to solo by a certified flight instructor. No signoff is required to fly an ultralight gyro, but it's still a good idea.)

Today's gyroplanes have other refinements and variations of the Bensen design. Some have different engines, different rotor blades and other different components, but most of them follow the basic Bensen Gyrocopter format.

Before Bensen

It may be comforting to know that the gyroplane has a history that goes back to Bensen's work during the 1950's. But even before that, the same basic type of aircraft had been flying since 1923.

It was only 20 years after the Wright brothers made their first airplane flight that the first gyroplane flew. The autogiro, an invention of Spanish nobleman Juan de la

Cierva, flew on exactly the same principle as today's gyros.

Later, the noted airplane designer Harold Pitcairn brought the autogiro to the United States under license from Cierva and began producing his own version of this aircraft. Pitcairn's very first model, the 1931 PCA-2, was a highly refined rotorcraft. It was so advanced that a PCA-2 autogiro that was restored in 1986 required no changes in the rotor system or any other parts of the aircraft. Before flying this restored machine, Harold Pitcairn's son Steven brushed up on his gyro flying skills in an Air & Space 18A gyroplane, a large factory-built machine that is basically an enlarged, closed-cabin version of a Bensen Gyrocopter.

Autogyros were refined and developed from 1923 through the early 1940's. Then a chain of events stifled public interest in these rotorcraft: 1) World War II broke out, drawing attention toward fast-flying, high performance airplanes; 2) The Germans demonstrated the first successful helicopter, the Focke-Achgelis F.61, and kindled a rivalry in the Allies to develop hovering machines; 3) The two foremost proponents of the autogyro, Cierva and Pitcairn, were killed in accidents, Cierva in an airplane crash, Pitcairn in a gun accident.

A 1931 Pitcairn PCA-2 autogyro like this one landed on the White House lawn.

34

The autogyro was highly developed when the Cierva and Pitcairn companies finally abandoned it. It could "jump" take off using no more space than a helicopter, it could fly fast or slow like a helicopter, and it could land in almost as little space as a helicopter. (A Pitcairn autogiro landed and took off from the White House lawn once!) But the one thing the autogyro could not do was hover, and the government, which controlled aircraft development during the war years with its purchasing power, was obsessed with hovering machines.

Even though a few recent attempts have been made to revive the autogyro (which was being called the gyroplane by this time), this type of aircraft has become almost entirely a sport machine. The German and British gyros of World War II were far from sporty, but they led directly to the gyroplanes you see in the sky today.

Your gyroplane has only twenty years less history than an airplane, and a very respectable pedigree!

35

Chapter 5

How a Gyroplane Works

When you look at a gyroplane closely, you discover that this seemingly random collection of parts is, in fact, a perfectly coordinated, stable aircraft.

There's a lot about a gyroplane to distinguish it from the more familiar shape of an airplane. But in fact, there's really only one fundamental difference: on a gyro the wing rotates; on an airplane the wing is fixed.

If you were to remove the wing from an airplane and substitute a set of rotor blades, you'd have the basic concept of a gyroplane. This, in fact, is how the earliest gyros were made: Juan de la Cierva built several of his early models using a fuselage taken from an Avro 504 airplane.

Rotor blades

The rotor blades are the wing of a gyro—a rotating wing. The rest of the aircraft hangs from that wing and serves to: 1) push the rotating wing forward through the air, and 2) point the wing in the direction the pilot wants to go—up or down, left or right.

The forward motion of the aircraft causes the wing to spin in "autorotation", meaning that the blades don't require direct engine power to keep spinning. Just by going forward 25 or 30 miles per hour, the rotor blades can spin fast enough to get their tips going 350 mph or so. That's why gyroplanes don't stall—because even at slow speeds their blades are still zinging through the air. That's also how a seemingly teeny six-to-eight-inch-wide rotor blade can hold up hundreds of pounds of gross weight. It's going fast enough to produce the required lift.

Spinning the blades that fast also makes them stiff, so that even highly flexible rotor blades can stand out straight and support the weight below them.

The engine

The job of the engine is simply to push the craft through the air. The engine is completely separate from the rotors of a gyroplane, and drives a separate propeller. When that prop pushes (or pulls), the machine moves through the air and the rotors autorotate from the forward motion. So the engine of a gyro has the same function as the engine of an airplane, to generate thrust that pushes the aircraft along.

The controls

There are three main controls: 1) the throttle, which controls the thrust produced by the engine, 2) the stick, which tilts the rotor blades and 3) the rudder, which points the nose left or right.

Since you've just heard about the engine, we'll move directly to the stick. There's a stick like this on airplanes and also on helicopters. To be correct, this stick is called the "cyclic" control. It tilts the disc formed by the spinning rotors. This is the primary control of the gyroplane, since it can make the machine bank and turn or raise and lower its nose. (The other stick, which only helicopters have, is called the "collective" control. It adjusts the pitch of each rotor blade. Nearly all gyros have their rotor blade pitch fixed, so no collective control is needed.) When an aircraft tilts left or right, it is said to "bank". When the nose moves up or down, that movement is called "pitch."

The rudder of a gyroplane works exactly like the rudder of an airplane. It makes the nose point left or right. This kind of movement is called "yaw."

You steer the rudder of a gyro or an airplane with your feet. Proper motions of the rudder pedals are very important during takeoffs and landings in a gyro, but are not so important in flight. After takeoff, you can make nice turns in a gyro with the stick alone and no change in the rudder.

In contrast, airplanes require the use of rudder to make "coordinated turns," since an airplane will try to move its nose left or right when you tilt the wing with the ailerons, an effect called "adverse yaw."

In either a gyro or an airplane, the rudder is used to cancel out torque or "P-factor" from the engine and propeller. Those terms refer to the tendency of the machine to yaw nose left or nose right when the engine is running fast and the aircraft is moving slowly, as on takeoff. The rudder handles this about as easily in a gyro as in an airplane since there's no torque from the freely spinning rotor blades.

The airframe

When you watch a gyroplane flying by, you're looking at one of the least important parts of the flying machine. You're looking at the airframe, which has about the same importance as the handle of an umbrella.

Aerodynamically, the purpose of the airframe is to help those invisibly spinning rotor blades do their work. The airframe hangs from the rotors, pulling them down against the oncoming air. And it points the movement of the machine in some particular direction.

But before you decide the airframe is so insignificant that it can be built of anything from water pipe to Legos, let me mention one other thing the airframe does: It carries *you*! The airframe has the very difficult job of carrying your lumpy, unstreamlined body through the air. It has to move you, a terrestrial mammal, in ways that won't upset thousands of years of adaptation by your earthbound ancestors.

That means the airframe must have control handles and pedals that you can work conveniently. It must have tail surfaces that will stabilize you comfortably. It must have wheels that effectively roll you on and off the ground (or floats to do the same thing on the water).

The airframe allows you to fly like Mary Poppins, hanging from your rotors the way Mary hung from her magic umbrella. Without her handle, Ms. Poppins could not have utilized her umbrella. Without an airframe, a set of rotor blades can't fly controllably.

You

Let's not overlook the part of your gyroplane you care about most: *you!* When you climb into that pilot's seat, you become an important part of the machine.

For example, on many machines you'll notice that when the pilot sits down, the tail wheel comes up and the craft rocks onto its nose wheel. That's because you, the pilot, are part of the weight and balance of the machine. Your weight is needed to make the gyro fly properly.

You are also an airfoil (not a very good one), unless you're flying in a completely enclosed machine. The air flowing around your body affects the stability of the aircraft. You may be surprised to learn that your body makes your aircraft more stable than the same gyro with an enclosure added. That's why extra tail surface is often added when an enclosure is put on a gyro.

But even though your body may be more of a stabilizing factor (because of its smaller size), it's less streamlined than most enclosures. So the enclosed gyro may be faster and more efficient without your carcass hanging out in the breeze!

Gyro designers take into account your size, shape and weight when developing open-frame machines. There are upper and lower limits to the size and weight of the pilot, who becomes part of the aircraft.

The parts of a gyro and what they do

Here's a quick review of the parts of a gyro, the ones you'll hear mentioned most. You may know all of these terms, or some of them, or none. But with this quick explanation, you'll be able to jump into any conversation about gyros and hold your own.

Rotor head - This is the device at the top of every rotorcraft that connects the rotor blades and the aircraft. In a gyroplane the rotor head allows the rotor blades to tilt fore and aft or side to side. It's connected to the control stick, which controls the tilting. The gyroplane rotor head is much simpler than that of a helicopter, since no power goes through the gyro rotor head and there's no need to change the pitch of the individual rotor blades.

There's some variety in the design of gyro rotor heads, but most are of the offset gimbal variety. That means the center of the main bearing is "offset" toward the rear of the machine behind the "gimbal" joints, the points where the rotor head tilts. This is done to make the machine more self-stabilizing so that, when properly trimmed, it tends to fly straight and level with hands off the controls.

Teeter bolt - This all-important bolt runs through the center of the rotor blades and the rotor head to connect the two. It's absolutely necessary for the rotor to be free to pivot on this bolt for flying stability. The rotor pivots as it spins to allow the advancing and retreating blades to provide equal lift on both sides of the machine. If the blades didn't pivot and were locked in place, the machine would roll toward the retreating blade and be uncontrollable.

In flight, the entire gyro hangs on the teeter bolt. That makes the quality of this bolt extremely important. A hardware store bolt just will not do as a teeter bolt. An aircraft-quality bolt must be used, which is free of flaws and surprisingly strong.

Torque tube - This is a part of many rotor heads which connects everything. It gets a lot of attention since it transfers all the control forces to the rotor blades and carries all the weight of the machine. It's actually not a tube on most rotor

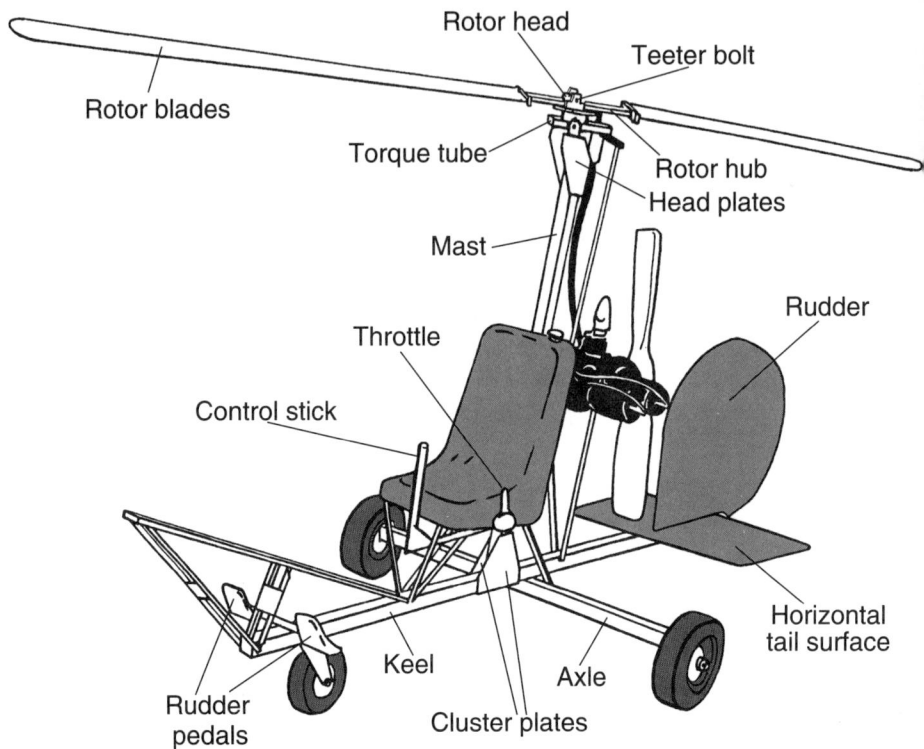

heads, since the original hollow tube used on early Bensen rotor heads was replaced by a solid bar by some builders. You'll find it just under the main bearing, parallel to the keel. Not every rotor head has a torque tube or uses this terminology.

Head plates - Many gyros attach the rotor heads to the mast between two metal plates. These are always made of carefully selected aircraft-grade aluminum or steel.

Mast - The mast is the upright piece of the airframe that usually sits vertically just behind the pilot. At its top is the rotor head. At its bottom is the rest of the airframe. Typically the pilot's seat and the engine are attached near the middle of the mast. This is also sometimes called the rotor pylon.

Keel - The front-to-back piece of the airframe is the keel. It's attached to the lower end of the mast and carries the rudder pedals and nose wheel at the front, the tail surfaces at the back.

Axle - This side-to-side piece of the airframe is connected to the bottom of the mast and to the keel. While it is commonly called an axle, it's actually not an axle but a cross brace that connects the real axles, on which the main wheels roll. In many machines an assembly of braces is used instead of a single axle piece.

Cluster plate - Many of the open-frame gyros use a pair of metal plates to connect the bottom of the mast and the keel. These plates are always made of carefully selected aircraft-grade aluminum or steel. Not all gyros have cluster plates.

Rotor blades - You already know a lot about these. They are the real aircraft; all the rest of the machine is there to make the rotor blades work properly. Nearly all small rotorcraft use two blades mounted exactly opposite each other. Each rotor blade is, in essence, an aircraft in itself, making a continuous circular flight around the rotor head. It must be balanced fore and aft (chordwise) just like an airplane. Each rotor blade also must be precisely balanced to match its mate and trimmed to track right behind the preceding blade.

Rotor hub - Every rotorcraft has some kind of rotor hub, whether it's a simple aluminum bar or a complicated assembly of parts. Its purpose is to hold the rotor blades in position, spinning with them and connecting them to the rotor head. On many gyros this is simply a very strong aircraft-grade aluminum bar with precisely machined fittings to attach it to the rotor blades and to the teeter bolt. Even this simple rotor hub must be made with great precision and is always produced by a factory-quality machine shop.

While it looks simple, the rotor hub actually does a couple of very complicated and important things: 1) It teeters, rocking with the rotor blades in a see-saw fashion. This

movement is critical to proper operation of the rotor blades, since it allows the advancing and retreating rotor blades to compensate for each other's different airspeed. 2) The rotor hub deals with lead and lag force, the force tending to wiggle each blade forward and backward as it goes around. Actually, the lead/lag force is typically resisted by the very solid rotor hub, and is handled by the resiliency of the mast and the rotor blades. The rotor hub is the connecting point for all this.

In machines with more than two rotor blades, the rotor hub and rotor head can get very complicated, with special lead/lag hinges. In helicopters the rotor hub and rotor head must also handle changes in the collective pitch of the rotor blades, as well as carrying engine power to spin the blades.

Control stick - This is one of three controls a pilot uses to fly a gyroplane. (The others are the rudder and the throttle.) The control stick steers the machine, banking left or right, nose up or nose down. It does this by tilting the rotor blades in relation to the airframe. That's different from an airplane, where the control stick activates small control surfaces (ailerons and elevator). The standard gyro control stick is a joystick similar to that used on airplanes. Some gyroplanes use an overhead stick which is connected to the rotor head directly over the pilot's head.

Throttle - Here's another of the three controls of a gyroplane. It functions like the accelerator of a car, increasing and decreasing engine power. But unlike a car, the affect of the throttle in a gyro is more complex, since changes in throttle affect altitude and trim of the machine.

Rudder - This third control makes the machine "yaw," swinging the nose to the left or right. On the ground the pilot may use the rudder to steer the machine with his feet much like a coaster wagon (with the control reversed so that pushing the right rudder pedal turns the machine to the right). In the air the rudder pedals don't have the primary job of steering. Instead they point the nose of the machine straight ahead (or some other direction at the pilot's discretion). Turning in flight is accomplished mainly by the control stick. The rudder of a gyro does not have to be used in flight for airplane-style coordinated turns, since there is no adverse yaw in a gyroplane.

Some gyros have a one-piece rudder that pivots completely. Others have a two-piece fin/rudder combination, with only the rear portion turning.

Horizontal tail surface - Some gyros have this, some don't. You can get into a lively discussion about whether this is necesssary for all gyroplanes. Many gyros have been flown successfully for years without a horizontal tail surface. The original Bensen gyrocopter, for example, had only a small horizontal "gravel shield" under the engine to keep rocks and debris out of the propeller. It had little if any effect as a horizontal tail surface.

Other gyros have a horizontal fin mounted far enough back to be effective at stabilizing the machine and capable of countering movement of the rotor blades, making control response more immediate. A horizontal tail surface is usually needed on a machine with a body enclosure to overcome the destabilizing effect of the body. On an open-frame machine, a horizontal surface is sometimes used, sometimes not.

Gyroplanes fill the sky in all sizes and shapes.

Chapter 6

Learning to Fly a Gyroplane

"It's one of the most exciting experiences of your life!" You've heard that before — about a movie someone wanted you to see, about some new sport your friend wanted you to try, about that blind date with your cousin's friend from Peoria. And no doubt you've found that most of those promises of fantastic excitement were exaggerated.

But, my friend, when I tell you to hold on to your hat and your heart when you start flying a gyroplane, you should believe it! This is not going to be like anything else you've done. It's going to be different. It's going to be fun. It's going to be thrilling!

Even the *thought* of learning to fly one of these open-air whirlybirds can increase your pulse rate a few RPM's. It's exciting enough to make you want to jump in and get started as fast as possible. But to be successful, you've got to get control of your enthusiasm, take a deep breath and give the process a little rational thought.

There are some important choices you have to make. Sure, you have to pick out the machine you want to own and fly. And chances are you've already discovered the stimulating variety of designs available from all those manufacturers. You can spend a lot of enjoyable hours looking through brochures, talking to people who own the machines that look best to you, and watching them fly.

Deciding on the particular aircraft design that you want is a very important first step, because it will influence how you go about learning to fly. Then there are other decisions you must make that are even more important to your

Paul Bergen Abbott enjoys a flight in his Gyrocopter.

success. At the top of the list is deciding what kind of flight training you'll get. This depends partly on what kind of machine you'll learn in and how you'll go about transitioning into your own gyro.

What kind of flight training should I get?

Let's take that first big decision: What kind of flight training to get. Notice that I didn't bring up the question of whether or not to have flight training. That is not a question; you *must* have flight training from a competent instructor to be successful in learning to fly a gyroplane.

You cannot teach yourself to fly a gyroplane. You cannot go it alone. Try it and you'll crash. That's not because this machine is so terribly difficult to fly. It's because you will undoubtedly make mistakes in the learning process. You may have learned to ride a bicycle without an instructor, but this is different. In learning to fly a gyro you can't afford to fall down, not even once!

To be successful, you need an experienced instructor on the machine with you to take you through a training sequence step by step. You need instruction regardless of your previous flying experience in other types of aircraft. This is true whether you have never flown in an aircraft or whether you pilot 747's every day around the globe.

When the U.S. Air Force assigned two experienced test pilots to learn to fly a gyroplane, they learned first in a dual-seat machine with an instructor at their side. Dual-seat training was needed by experienced military test pilots. You need it, too.

The first question is not whether to have training, but how to get it. Today there are engine-powered dual-seat training gyros that are specially designed to fly as much as possible like your single-seat machine. You can take lessons in one of these two-seaters in comfort and safety, with a skilled instructor next to you. This makes learning to fly a gyroplane as easy as learning to fly an airplane.

In the past, the only way to learn was in a two-seat unpowered gyro that was towed behind a car or truck. While that method is still available today, the engine-powered trainer is by far the best way to learn.

Learning to fly in an engine-powered trainer

The engine-powered gyro trainers give you a flying experience that's more like flying your own single-seat machine. In fact, the powered trainers are essentially identical to the single-seat machines except for the extra person aboard and the heavier total weight.

You can climb aboard a powered trainer, take off and decide rather quickly what you think about flying in a gyro.

Two seats and an instructor beside you make learning to fly a gyroplane fun.

The special feel of these light, breezy rotorcraft is something that pleases a lot of people — and disappoints others. It's nice to be able to find out your own particular reaction without first having to pay to buy a machine.

Some instructors in powered trainers will quickly take you up several hundred feet high, giving you a quick look at the world from that perspective. Others will work you up gradually, giving you a chance to get used to being in the air at low altitude before pulling the ground out from under you. Either way, you're going to find out very soon what gyro flying is like in a powered trainer.

Another big advantage of the powered trainer is that the student gets experience with all the controls he will use in his own single seat machine. He will become proficient in handling the control stick, the rudders and the throttle. In the towline gyroglider, the rudders and the throttle aren't there, and they have to be learned later in a single-seat machine.

The powered trainer will take off, fly and land just the way a single-seat machine will do. The airspeeds for liftoff, for maneuvering and for landing will be about the same. On takeoff the powered machine will lift off about the way a single-seater will.

These powered dual-seat trainers are experimental aircraft, built with the same kind of semi-rigid rotor system that's common on homebuilt gyros. They fly at similar speeds and handle very much like single-seat homebuilt gyros. While some of these trainers shield the student and pilot comfortably in an enclosure, others are open-frame machines that will demonstrate the feel of this type of open-air flying.

The powered trainer also allows the student to bypass the hardest part of gyro flying, the takeoff, and go right on up to practicing the easier art of straight and level flight. The instructor can do the takeoff; then, with everything nice and level, he can hand off the control to the student.

The student's first experience in a powered trainer can be the comfortable condition of a level, stable flight attitude. There, in straight and level flight, he will typically discover that flying a gyro is not a superhuman challenge after all, and his confidence will go up. This can make it seem so easy that it's possible for a student to underestimate the skill required to fly one of these exotic birds. He may go off to try flying on his own before he is ready. The instructor, not the student, must decide when the student is ready to solo.

The Parsons Trainer

There are several machines that are used for powered dual-seat training. If you're interested in an open-frame gyro, you may prefer one of the open-frame trainers that will closely resemble your machine. One of these is the Parsons Trainer developed by Bill Parsons of Lake Monroe, Florida. It closely resembles the Bensen gyrocopter, but seats two people in a tandem configuration, one in front of the other. In this machine, the student can sit out in front for a very realistic view of the world as it looks from a gyro seat.

The RAF 2000 side-by-side gyroplane

For a different approach, another option is a gyro which carries two people side by side. Some people prefer this seating arrangement; it may seem a little more friendly having your instructor at your elbow. Some examples of this type are the Air Command side-by-side two-place and the RAF gyros.

Besides these experimental trainers, other gyros have been used successfully for flight training. Along with homebuilt gyros, there are two factory-built two-place gyroplanes, the Air and Space 18-A and the McCulloch J-2.

Air and Space 18A

These two big machines have rotor systems that are different from homebuilt gyros, but they are acceptable by the FAA for training.

Getting started with training in a powered machine doesn't require any paperwork or federal license. If you plan to fly without a pilot license in an ultralight gyro, you will never need to bother with paperwork. But if you want this training to apply toward a pilot's license,

a federally-licensed Certified Flight Instructor (CFI) can record your training in a log book.

Is it really worthwhile to go to an instructor to get training in a dual-seat powered gyro? It will take some time, some money and some

McCulloch J-2

bother. Unless you are one of the lucky people with a gyroplane instructor nearby, you may have to travel many miles to find someone who gives powered gyro training. But consider this: Your own gyro is going to cost you thousands of dollars. The cost of traveling to a powered training center for lessons can be measured in hundreds. For most people it's cheap insurance (and lots of fun) to go ahead and shell out the money needed for powered training.

Mind you, you won't find powered gyro schools in the Yellow Pages of every town. Powered training gyros came into use only recently, starting about 1985, and they are now available in several locations around the country. Unless you're lucky enough to happen to live near a powered gyro training center, you'll probably have to travel to it.

Some instructors offer a training package for a set price, guaranteeing to give you a certain number of hours of instruction and promising that you'll solo. Some offer a weekend package of concentrated training. In either case, make sure you have enough training to be confident about going home to begin flying your own machine, and make sure your instructor agrees that you are ready.

Learning to fly in an unpowered gyroglider

Long before there were powered two-seat trainers, gyro pilots all over the world were learning to fly by a different method: the towline gyroglider. Soon after Igor Bensen developed the first gyrocopter in the mid 1950's, he recognized that he'd have to come up with a method for teaching people to fly his unique aircraft.

Towed by a car, an engineless gyro makes a useful trainer.

Paul Bergen Abbott enjoys flight on a towline.

The method Bensen developed was the one he himself had used to test out his machines. He simply pulled an engineless version of his aircraft on a rope behind a car. Using careful coordination with the driver, it was possible to simulate powered flight and practice handling the rotors and the control stick, as well as getting some idea of how the machine felt in flight.

Once towline flight was mastered, the student was then allowed to climb into a single-seat powered machine and begin to work up to flight very gradually. I emphasize gradually because this is very risky business, and moving slowly is necessary to make sure the student's inevitable mistakes are small ones. The general sequence for transitioning to power was: slow taxi practice without the rotor blades mounted, gradually faster taxi practice with rotor blades spinning, short tentative hops off the ground, longer hops, straight-line flight down the field, S-turns down the field, up-and-down flight down the field and finally a higher-altitude first flight around the field.

To accomplish this training, Bensen set up his own flight school and successfully taught many people to fly

Gyrocopters. He called his two-seat trainer a "gyroglider." It was simply a Gyrocopter without an engine and with a double-wide seat.

Besides being good training, this towline gyroglider technique was a lot of fun. Igor Bensen gave training this way during the early years when there were no other gyro instructors. Eventually other volunteer instructors built gyrogliders and joined in the training.

Today the towline gyroglider method of training is still available in some locations. It can still be an effective way to start your training and is usually less expensive than learning in an engine-powered machine. Some local chapters of the Popular Rotorcraft Association (PRA) operate towline gyrogliders for their members.

These PRA club trainers can be successful when they are used for initial flight training. This is a low-cost and relatively simple way to give new pilots the feel of the rotors and a sense of how a gyro operates. However, this initial training on a towline must be followed by dual-seat training in a powered machine.

Training skills

Whether you start your training in a powered trainer or a towline gyroglider, you will develop certain basic skills. To help you anticipate them, we'll list them and give you a brief description. This is not enough information for you to learn these skills on your own It takes more than information. It takes instruction and experience. This list will help you understand what's ahead of you.

Bringing the rotor blades up to speed - With or without a pre-rotator, you learn to do this gradually and smoothly. You learn how to avoid flapping (a bumping feel in the stick caused by going forward too fast for your rotor speed). If not recognized, flapping can damage your rotor head or rotor blades and make them dangerous to fly.

Taking off - In a powered trainer you take off just as you'll do in your own single-seat machine. You lift off gradually, flying away from the ground with the nose raised just enough to begin flying. This is one of the advantages of the powered trainer, since it accurately duplicates the takeoff performance of a single-seat gyro. But a towline glider takes off differently: On the towline you typically take off with the rotors all the way back (full BACK stick) so that you leave the ground at the lowest possible speed. As soon as you leave the ground you neutralize the stick and level out the rotors.

Developing reflexes on the control stick - Your control movements of the stick become automatic. You no longer have to think about which way to move it. It becomes a reflex, like steering a bicycle. You learn the great sensitivity of the stick, particularly from side to side, and you discover that the sensitivity increases at higher speeds. You learn how to handle the stick gently, gripping it lightly in your hand.

Making turns - You learn how it feels to bank in a turn. You learn to make "S"-shaped turns down the field at gradually increasing angles of bank. Whether you realize it or not, you are learning that as you increase your angle of bank you have to add a bit of BACK stick to hold your altitude.

Going up and down - You practice climbing and descending, making smooth hill-shaped patterns. In the process you learn what it's like to get up off the ground, and how it feels to be in the air (In a gyro 30 feet of altitude feels like 300!).

Avoiding PIO - You become familiar with the handling characteristics of the gyroplane. You learn that a gyroplane is uniquely responsive to the controls. Your control stick tilts the entire lifting surface, not just part of it like the aileron on an airplane wing. You learn that when you move the stick to bank or lift the nose, you make another movement of the stick to stop the change. You learn to damp your movements by reflex, because it just "feels right." This prevents overcontrolling with increasingly severe stick movements, which can lead to an increasing nose-up nose-down movement called pilot induced oscillation (PIO) or "porpoising." PIO can be disastrous, but you learn to avoid it completely through lots of flying practice with your instructor.

Landing - Both towline and powered trainers land the same way. You fly down to about a foot or two above the ground, and as you slow down you gradually increase UP stick until eventually the machine settles down gently, touching the tailwheel first.

Ground handling - You learn when to keep your rotors parallel to the ground while taxiing around corners (full FORWARD stick) to keep the rotors from striking objects on the ground. You learn to make turns on the ground only when the machine has slowed down.

Experience in the air

As you learn to fly in a powered trainer, you quickly find out how things look and feel from the air when you're up and away from the ground. For those who start in a towline gyroglider, this experience is missing. It's important to develop a level of comfort in the air, so you won't be distracted in your early solo flights by that lofty view of the world.

As you begin your gyroplane training, it's very helpful to get some air experience, either with some flights in a powered gyroplane or with some flight time in a light airplane. It helps to find out in advance how it feels to leave the ground and just keep going up, high above the ground.

You can take a ride in a light plane at your local airport. Pick an airport that gives flight instruction (Almost all of them do.), walk into the office and say, "How much for a flying lesson?" They'll quote you a price and will be eager to schedule you at the first opening. There's no paperwork or license required and they will put you in the pilot's, not the passenger's, seat. Some flight schools offer a special low-priced introductory lesson that will help keep the cost reasonable.

To get started in your gyro flight training, you don't need any kind of a license and you don't have to have your machine bought or built. All you need to do is find a gyro flight training center or a rotorcraft club chapter that you can get to, and get on their training schedule.

One of the most exciting experiences of your life is about to begin!

Chapter 7

Equipment You'll Need

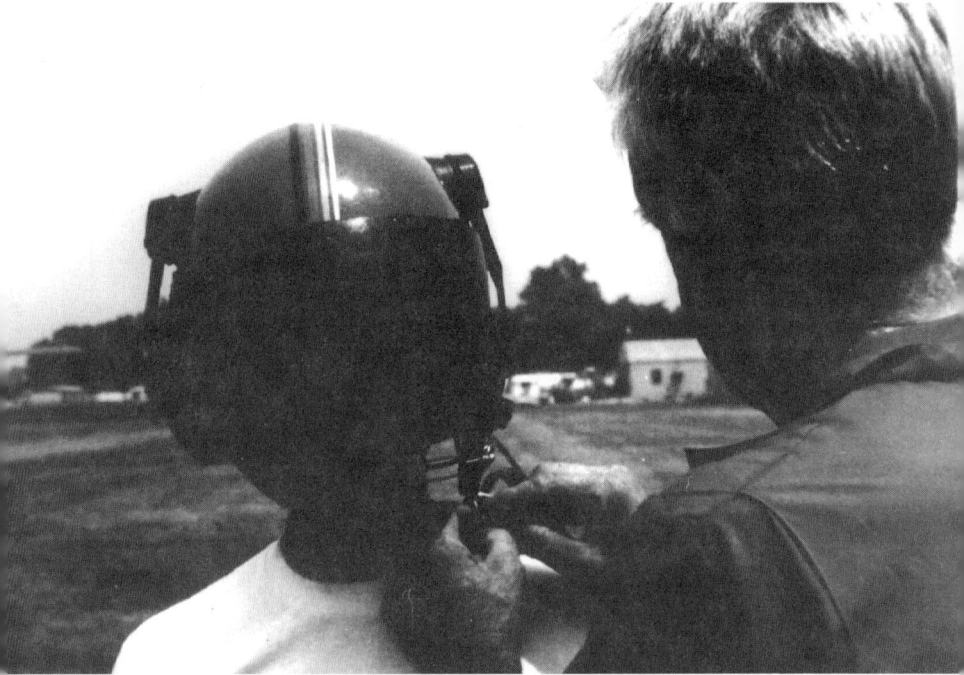

There isn't much equipment you need for flying your gyroplane, but the equipment you do need is extremely important. You need to have protection for your head and your body in case of an accident. That protective gear will also make you more comfortable, since it will shield you from the wind and chill you get blasting along a mile a minute several hundred feet above the ground.

The Most Important Hat You'll Wear

Without a doubt the most important piece of equipment you'll need is a good safety helmet. A good-looking helmet will do great things for your image as a modernistic experimental aviator. It will also do something even more important: A good helmet will protect your most vulnerable piece of anatomy from serious jolts in the event of an accident. A blow that might only raise a welt on your arm or leg could kill you if it hit your unprotected head.

The right safety helmet could save your life in even a minor accident. But to do so, it has to be built for something more than snappy appearance. A shiny blue metalflake finish with racing stripes does nothing to protect you. But luckily, some of the best-protecting helmets are also some of the best-looking. You can have your protection and blue metalflake, too!

How a Helmet Protects You

A helmet protects your head in two ways: The hard outer shell resists penetration and abrasion and absorbs the large initial shock in an accident. The softer inner liner absorbs the rest of the shock by slowly collapsing under the impact. Both the shell and the liner essentially self destruct by spreading the forces of the impact throughout the helmet material. That's why a helmet that survives an accident may not provide much protection afterwards.

A helmet damaged in an accident should be replaced. Some manufacturers will inspect a damaged helmet and may also be able to repair it. Note that a helmet can also be damaged by rough handling, such as being dropped sharply onto a hard surface.

How Helmets Are Made

There are five primary parts of a safety helmet:

1. Outer shell

2. Shock-absorbing liner

3. Padding

4. Chin strap

5. Face shield

The **outer shell** is usually made of either fiberglass or injection-molded plastic. The injection-molded plastic helmets (of "polycarbonate" material) can be damaged by storage near gasoline, cleaning fluids or exhaust fumes. They may also be damaged by paint or decals. The fiberglass helmets (actually they're usually polyester resin reinforced with glass-fiber cloth) are generally more expensive. The fiberglass shell absorbs shock by delaminating under impact. You can paint these helmets or apply decals without damage.

A **liner** for absorbing shock is made of expanded polystyrene foam ("Styrofoam"). In an accident, the liner cushions most of the impact. But in doing so, it's usually damaged, crushing under the impact. After that happens the helmet's full protective value is lost. Fortunately, some manufacturers will replace a liner that has been mashed.

Padding is made of cloth and soft foam strategically placed over the shell's inner liner to orient the helmet properly on the head. The padding does not cushion impact; It's too soft for that. What it does is make the helmet fit comfortably.

A **chin strap** is absolutely necessary, since a helmet will do you no good if it comes off during an accident. Naturally, the chin strap must be fastened.

A **face shield** may be built into the helmet or attached to it. It protects your face from the blast and chill of the wind when you fly, and it keeps away flying debris when you crash. A face shield is really the only thing about a safety helmet that's optional; You could wear goggles instead. Even if you already wear glasses, you need good protection over your eyes to keep out the bugs and grit and to prevent your eyes from watering and blurring your vision at a critical moment.

Types of Safety Helmets

There are three basic types of helmets available for gyroplane flying: 1) The "full-face" helmet encloses the head and neck and has a built-in chin section and face shield. It's the best. 2) The "three-quarter" (sometimes confusingly called "full coverage") helmet has no chin piece and uses a snap-on face shield. Some pilots find this type of helmet more comfortable and claim the feel of the wind on their face is the best airspeed indicator. 3) Partial-coverage (sometimes called "ultralight" helmets) leave the back of the neck unprotected as well as the face. They are the lightest and most comfortable, but are of questionable value in flying a powerful gyroplane.

The Best Helmets

If you can accept the experience of motorcycle riders, it has been found by researchers at the University of Southern California (USC) that the full-face helmets did protect better against face and eye injury in actual traffic accidents. But the biggest factor was the use of some kind of helmet *vs.* no helmet at all. With a helmet, the chances of surviving a motorcycle accident were considerably greater than without head protection. There's a lot of similarity between crashes of motorcycles and crashes of gyroplanes. With both vehicles, there are people who would have stood up and walked away from crashes if they had been wearing helmets, but who died instead of head injuries.

There are three major organizations which have established standards for the protective power of helmets. Each helmet that meets these standards will be marked with the appropriate seal.

DOT The Department of Transportation (DOT) label, found on the outside back edge or a helmet, means that the helmet is certified by its maker to comply with federal safety standards for helmets. Since 1980 all adult-sized helmets sold for motorcycle use have been required to meet DOT standards. One of these requirements is that the helmet must reduce the test impact to no more than 400 G's (That's 400 times the normal pull of gravity, and that's the kind of smack a head can take in an accident!). You should consider a DOT rating to be a basic requirement for a helmet for gyroplane use.

ANSI The American National Standards Institute (ANSI) has a more stringent set of standards than DOT, in the ANSI Z90.1 requirements. For example, the Z90.1b-1979 standard reduces the test impact to 300 G's, 75% of the DOT standards. ANSI represents certain consumer groups, helmet manufacturers, testing organizations and the military. The ANSI sticker is usually located on the inside of the helmet.

PHOTO BY CRISTINA ABBOTT

Put on a good helmet to look sharp and be safe.

Snell The Snell sticker represents the highest protective standards in helmets. This sticker is usually found on the inside lower back of the helmet. The Snell standards have been revised many times since the organization began testing helmets in the 1950's. Revisions are made every five years—in 1985, 1990, *etc.*—each time setting a tougher standard. The Snell Memorial Foundation is an independent organization founded in the memory of a man killed by head injuries. Snell scientists pound, puncture and crush hundreds of helmets to find which ones work best.

Note that DOT, ANSI and Snell are not brands of helmets. They are seals of approval of helmets that meet certain protective standards. These seals do not appear on all helmets, only those sold for motorcycle use. The seals probably will not show up on military helmets or helmets made exclusively for aircraft use. There are many companies who manufacture helmets carrying these seals of approval, including AMF, Arai, Bell, Fulmer, Honda, Nava, Shoei, Simpson and Yamaha.

Are There Bad Helmets?

We've already said that a helmet is no good if it has been damaged in an accident or a severe drop onto a hard surface. But there are other things that can make a helmet unacceptable. You should not drill holes in the shell (Install your intercom without making holes!). Also, a helmet can get just plain worn out from use: The shell gets cracked, the liner gets mashed, the chin strap gets frayed, etc. This kind of wear is accelerated if you hang your helmet on things, where the fragile inner liner can get dented.

Helmet manufacturers recommend that you throw out your old helmet every few years and buy a new one. Naturally, that's good for the manufacturers, but they say it's also good for you, for two reasons: 1) An old helmet may have lost some of its protective capability, and 2) Since improvements are constantly being made, you'll have the benefit of the latest developments.

Considering the high cost of a new helmet, I'd rather take really good care of my old one, thank you. I suggest you do the same. If you're not sure just how old your helmet is, look at the chin strap. It will have the month and date of manufacture stamped on it if it was made since 1974.

Getting a Helmet that Fits Right

Don't judge a full-face helmet the minute you put it on. At first, these plastic cocoons feel different from any headgear you ever wore. They fit tightly, so much so that you can have a bit of claustrophobia at first. You may even feel that you can't breathe but, believe me, you can. You may feel that it restricts your vision, but it doesn't affect your side vision. (Most people can see at an angle of 110° to 115° to the side, but helmets must be open 120°.)

A full-face helmet should fit quite snugly. (It may even feel a bit too tight until it's placed correctly.) You need to go to the store and spend a little time trying on full-face helmets and walking around until you adjust to them. You're not likely to buy a good-fitting full-face helmet by mail order, since hat size is not a reliable indicator of helmet size.

To fit, the cheek pads should touch your cheeks without pressing too hard, there should be no gaps at your temples and if the helmet has a neck roll, it should not push the helmet away from the back of your neck.

If the helmet fits just right, your skin will move when you move the helmet with your hands from side to side and up and down. You should feel as if a slight, even pressure is being exerted all over your head by the helmet. Remember that until a new helmet is broken in, it should fit as tightly as you can comfortably wear it.

Other Protective Gear

If all this talk about helmets gives you the impression that they're important, you're right. A helmet rivals your seat belt as the most important piece of safety gear you have on your gyroplane.

There's other equipment that can protect you. For your early flights, it would be smart to wear a stout jacket, preferably leather, to protect you from abrasion or flying parts if you tip over. Wear heavy pants, like jeans, for the same reason. This type of heavy clothing will also reduce the blast of the air against your body on an open-frame machine, making it more comfortable on all but the hottest of days.

Where to Fly

Besides protective gear, you also need a good location for your first powered solo practice. For your first taxi runs, a hard surface will be more comfortable. Later, when you're ready to do some higher speed runs and take off, a grass surface is better. While grass is bumpier, it's more forgiving of sloppiness in landing. Grass is also more forgiving of accidents, since the turf tends to absorb some of the rotor blade impact if you tip over.

To leave the ground you need a strip long enough to allow you to take off, do at least a couple of maneuvers and land again, all in a straight line. That takes 3,000 feet or more—the more the better. There should be no tall obstructions at either end to get in your way if you want to fly around the field at a fairly low altitude. Considering your first solo passes around the field, the more area under you that will permit a landing, the better. If you can't make the whole circuit over friendly territory, you will have to be confident in the ability of yourself and your machine to stay in the air until you're back over a good landing area.

Choosing the best location for your solo flights is one of the most important factors in learning quickly and safely. You need lots of space in which to take off, to land, and to learn.

Chapter 8

Mental Preparation

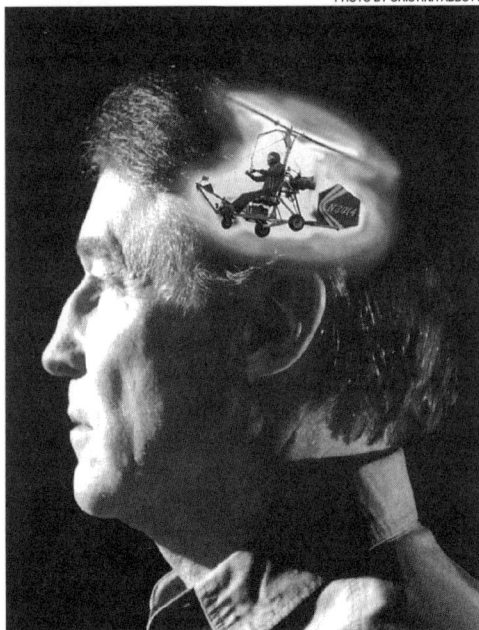

PHOTO BY CRISTINA ABBOTT

The idea may seem strange at first, but flying a gyroplane is all in your mind. After all, your machine already is capable of flying. Your muscles are already fully capable of every movement that's needed. As soon as your mind catches up and learns how to cope with this new situation, you'll be flying. So it stands to reason that mental preparation can help you get off the ground quicker and more easily.

One way you can prepare mentally is in dealing with fear. While many people won't admit to any fear in flying a gyroplane (or perhaps don't remember it), everyone experiences it. As one experienced gyro pilot said, "Even after years of flying a gyro and hundreds of hours in the air, I'm always just a little bit scared up there." By recognizing the sources of your fear and dealing with them, you can keep fear from getting in the way of your learning.

There are three main sources of fear in gyro flying: 1) noise, 2) speed and 3) height. All of these can be handled through mental preparation.

Noise

Let's face it. When you're sitting just a few inches in front of the engine of a gyro, it's plenty loud! Even with a good muffler system it's still enough of a racket to shake up your composure. You can hear the power of that fuel exploding in those cylinders, and you can feel the turbulence of that propeller thrashing violently against the air. Unfortunately, this is just the kind of thing that touches off one of our basic instincts, which is to associate noise with danger.

From the moment we are born we instinctively fear loud noises. Overcome this instinct and the fear will disappear. To do this you have to realize that gyroplane engine noise does not mean danger. Far from it. That singing engine is your safety signal, meaning that all is well. It's when you hear that sudden silence that you need to start worrying!

You are used to being close to engines that are considerably more powerful, like the engine in the automobile you ride in so nonchalantly. That auto engine is actually making as fierce a racket as a gyro engine, but heavy mufflers and soundproofing keep the noise from your ears. Instead of thrashing the air with a propeller, the output of an auto engine is buried in a transmission where you don't hear it. Without the noise you don't feel fear in a car. So why should you feel fear when the noise of your gyro is there? It's instinct, not logic. Protect your ears with ear plugs and get used to the sound of your gyro engine. Hang around gyros and listen to them run. Say to yourself,

"NOISE DOES NOT MEAN DANGER.
NOISE DOES NOT MEAN DANGER."

Speed

A second cause of fear is speed. Sitting so close to the ground, the sensation of speed in a gyro is exaggerated. Without a windshield, the rush of air is amazingly strong and contributes to the effect. It makes the speed seem far greater than it really is. Fifty miles an hour in a gyro seems like a hundred!

You can talk to yourself to anticipate this sensation of speed, but you can also experience it in advance. Beg, borrow or rent a motorcycle. Put on your flying helmet and face shield and spend some time riding at 50 miles per hour, which is about cruising speed for a gyro. You will soon get used to the sensations so that they won't seem exaggerated when you are on the gyro.

Another way to make the speed less bothersome is to consider that when it comes to flying, airspeed is a safety factor. Airspeed keeps you flying and gives you control. This is just the reverse of our experience with ground vehicles, where the danger *increases* with speed. In flying, the general rule is that danger *decreases* with speed (up to your cruising speed). Keep thinking,

"AIRSPEED IS GOOD. AIRSPEED IS GOOD."

Height

A third cause of fear is height. In a gyro a little altitude seems like a lot. That's one good reason for learning to fly in a powered two-seat machine. You'll get used to the feel of altitude, and you'll learn to concentrate on your flying instead of the height. As an alternative, it's helpful to get some experience flying in a light plane, where you can also see how things look from up off the ground.

Besides dealing with fear, there are other mental tricks that will allow you to learn to fly more quickly. There are three kinds of *simulation* you can use. The first is *mental practice*. This simply means thinking through all the sensations and movements of flying while you are on the ground. You can do this anywhere, on a bus, at a desk, in a chair. You simply close your eyes and visualize yourself in the seat of a gyro. Then use your imagination to think through every detail of a flight from start to finish. Imagine starting up the rotor blades, taxiing to the end of the field and taking off. Imagine every control movement. Imagine making gentle

The view from the seat of an open-frame gyro takes mental adjustment, even with an instructor beside you.

banking turns in each direction. Imagine getting hit by a gust of wind, and any other situation you think you might encounter.

These mental practice sessions build up reflex patterns very much like actual performance. It's a technique used by professionals in complex skill events like golfing, bowling or pole vaulting. Watch Arnold Palmer as he lines up a putt. He's thinking through the shot, practicing it in his mind. Watch a pole vaulter as he stands waiting at the end of the runway. Why does he wait? Simply to do one last mental practice jump before starting on the real one.

One step up from mental practice is *static practice*. This means sitting in your machine while it's parked in your yard or garage and actually making the control movements of

71

a complete flight. This works just like mental practice except that you are giving your mind stronger signals by reinforcing them with muscle movements. You can even add realism by making engine noises with your lips, so long as nobody catches you!

A third type of simulation is *rehashing*. This is a way of getting greater training value out of your actual flying practice. After each flying session, sit quietly and mentally re-create the last trips down the field. Include every movement and detail you can remember, thinking about what you could have done differently to improve your performance. This has the mental effect of additional trips down the field, and can show you the solutions to problems you run into.

Chapter 9

Flying Characteristics of the Gyroplane

Anyone can tell just by looking at a gyroplane that this type of aircraft is different. It obviously doesn't look like an airplane, and it only looks a little like a helicopter.

But how does a gyroplane fly? Are the flying characteristics very different from an airplane? from a helicopter?

Surprisingly, some people have ignored the unique appearance of gyroplanes and decided that they must fly more or less like an airplane. One such person was a fixed wing pilot with a Private Pilot license and quite a few hours in his own Cessna 150. Assuming a gyro would fly just like an airplane, he refused gyro flight training and took off in his newly-built gyroplane. Naturally, he crashed. The amazing thing is that he rebuilt his machine and, still convinced the gyro had to fly like a Cessna, he repeated the whole story. Takeoff again. Crash again!

To avoid any confusion, let's be very direct about it: The gyroplane does not fly like an airplane. Yes, there are some similarities, but the gyro has some flight characteristics of its own. And even though it looks a lot like a helicopter, the gyro flies like a gyro.

This chapter is here to help you understand the unique characteristics of this type of aircraft. To avoid talking about a particular make and model, we're going to talk about the flight characteristics of the gyroplane. The term "gyroplane" refers to any unpowered rotor flying machine.

The gyroplane is not a particularly difficult aircraft to fly. It's certainly much easier to fly than a helicopter, and in many ways it's easier to fly than an airplane. The people who find the gyroplane easy to fly are usually those who understand the unique flying characteristics of this type of aircraft. They find that once the gyroplane is understood, it really is easy to fly.

So to help you understand this machine more easily, here's a discussion of the main features of the gyroplane:

Advantages of the gyroplane

The gyroplane was invented for one particular advantage: to eliminate the danger of a stall. It was developed by Juan de la Cierva, from Spain, after the new airplane he designed was destroyed when it stalled and crashed. Cierva set out to create an aircraft which was incapable of stalling. In only four tries, he completely succeeded. Cierva not only invented rotorcraft, he invented the only characteristically stall-proof aircraft to date. His design, which he called the "autogiro," was the first gyroplane. Over the years the gyroplane has been developed and refined even further.

In the decade of the 1950's, designer Igor Bensen changed the handling characteristics of gyroplanes with his short-coupled pusher-engine design, which he trademarked the Gyrocopter. All modern gyroplanes have followed the configuration Bensen used. Since the earliest years, all gyroplanes share one magnificent characteristic: *They will not stall!*

1. Gyroplanes won't stall - If you were to fly an airplane slower and slower, you would become very concerned about a stall. At a slow enough speed the wing will simply stop flying. The airplane will fall until it regains enough speed to fly, or until it hits something. If an airplane stalls near the ground, there's not much chance of survival.

The gyroplane, however, can't stall and make this deadly plunge. No matter how slowly you fly it, you keep control. And rather than a wild nose dive, the gyroplane will simply begin to settle and lose altitude at slow speed. Properly handled with an application of power, the gyroplane can always regain flying speed or can be landed.

**With no hands on the controls, Paul Bergen Abbott demonstrates
the stability of the gyroplane.**

2. Gyroplanes won't spin - A spin happens when one wing
of an airplane stalls while the other keeps flying. Since the
gyroplane doesn't stall, it can't spin.

If a gyroplane won't stall or spin, then isn't it foolproof?
Could such an aircraft ever be dangerous? The answer has to
be 'Yes', since gyroplanes can be dangerous just as automobiles are dangerous. Gyroplanes and automobiles are powerful vehicles that can go 50, 60, even 100 miles per hour. But
the difference is that gyroplanes have a better safety record
than automobiles. They even have recorded a better safety
record than factory-built airplanes. You can read more about
this in Chapter 14.

You're quite willing to step into an automobile and drive
away without ever thinking about the danger. Why?

Because you're in control and you're managing the risk. You don't expect to have an automobile accident. Likewise, the gyroplane, in spite of its unconventional appearance and exuberant engine, can be managed so that the risk is minimal.

To manage the risk of flying a gyro, you need to be thoroughly familiar with the limitations that are unique to this category of aircraft. Five of these limitations are listed below and suggested remedies are discussed.

This enclosed Air Command 147A flies well.

Limitations of the gyroplane

1. Damage during ground handling - It's possible to damage the rotor head and the blades by improper handling on the ground. Flapping (mentioned earlier in Chapter 6) can stress and fatigue a rotor head if it's not routinely avoided. And if the pilot isn't alert while taxiing, the blades can hit rocks, runway lights, and the like.

> **Remedy:** Lots of takeoff practice is needed, learning to avoid flapping while taxiing up to flying speed. Also keep the rotor blades flat to the ground (full FORWARD stick) when turning or taxiing near obstacles. Carefully inspect the rotor head and blades before every flight.

2. Getting behind the power curve - This may be the most common problem among beginning gyro pilots. It's really a very simple idea, though people usually explain it with a graph, drawing a curve from which the problem gets its name. It simply means allowing the airspeed to get so slow that the engine can't maintain altitude, even at full throttle. This is a vital characteristic to understand. Here are the main situations where it can happen:

a. At takeoff - Here's an actual example: A new gyroplane pilot had been making powered flights down the airport runway at 20 to 30 feet of altitude flying straight and level and making shallow "S" turns. He decided he was ready to make his first trip around the pattern. After flying past the end of the runway, he found his machine wouldn't climb. He squeezed under some highway power lines, buzzed through a back yard at three or four feet of altitude, and finally made it back to the field, a nervous wreck!

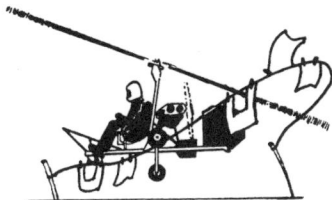

Remedy - The symptoms indicate he was flying "behind the power curve." His airspeed was too slow and he should not have attempted to climb. He appeared to be flying nose high, preventing his airspeed from increasing, even at full throttle. In this case, the problem was remedied by changing the trim of his offset gimbal rotor head. He raised the tension spring clamp at the top of his mast. This made it feel natural to lower the nose in flight and his ability to climb became normal. (Another cause might have been lack of thrust from poor engine performance or an improper propeller, either of which would show up in a static thrust test. Other possible causes might have been the decreased lift of a hot and humid day, or a machine that was overweight or didn't pass its hang test.)

b. During level flight - As long as you maintain cruising airspeed, you shouldn't get behind the power curve during level flight. But watch out for *downwind turns*. This is a common place for problems. I've seen it happen: The pilot was flying upwind about 25 feet high. There were lots of people at the flying field and he was concentrating on making a good-looking turn downwind. Instead of watching his airspeed, he carefully watched the ground, maintaining a smooth angle of bank and a constant groundspeed. Suddenly he found himself settling to the ground. He gave the engine full throttle but still continued to settle. He had no choice but to land, fortunately in a clear area.

Remedy: Don't confuse airspeed with ground speed, especially during downwind turns. Your ground speed will have to actually increase, while your airspeed will stay the same. Keep your airspeed constantly at the proper cruising speed whatever the maneuver, and you'll stay in the air. Also, it helps to have a little altitude before making a downwind turn since this makes changes in ground speed less noticeable. Every turn requires more power, or else the craft will lose altitude or airspeed, or both.

c. During landing - I've never heard of a serious power curve problem on landing. However, it's worth mentioning that you should fly down to within a few feet of the ground at full cruising speed. Only then should you reduce speed and flare. If you were to slow down before you were near the ground, you could get behind the power curve. Then, if you didn't have enough altitude to drop the nose and regain some airspeed, you would probably land short of your intended spot and a bit rougher than you would prefer.

3. Rotor blade stall - Gyroplanes don't stall. They can't. But at very high speeds their rotor blades can. It takes something like 150 mph, according to designer Igor Bensen. That makes rotor blade stall almost only a theoretical possibility, since it would take an extreme maneuver like a power dive to produce such a heroic speed as 150 mph in a gyro!

What happens? At a very high forward speed, as each blade spins around to the left side and moves toward the rear of the machine, it stalls momentarily and loses lift. Note that both blades don't stall; only the left, or "retreating" blade stalls. And only a portion of the lift is lost, usually from only a portion of the left blade. The nose does not drop as in an airplane. There's just a tendency to roll to the left, which can be anything from subtle to severe. Generally, a proficient pilot can maintain control.

You can feel the retreating blade stall as a distinctive two-per-rev (two bumps per each revolution of the rotors) vibration in the control stick. You won't feel it at normal flying speed, since the lift will be equalized on both sides of the machine.

Remedy: If you don't hear much about rotor blade stall in gyroplanes, it is probably because experienced pilots don't seem to worry about it much as a serious possibility. Igor Bensen has said that rotor blade stall "...is very seldom encountered by rank-and-file gyrocopter pilots and is usually known only to test pilots and engineers."

Experienced pilots never approach the kind of super speed required to stall the retreating rotor blades. As a final precaution, they're alert for the tell-tale two-per-rev vibration that occurs when rotor blade stall starts to happen. This feels different from the one-per-rev shake of out-of-track or out-of-balance rotors. Two-per-rev is a warning given by rotor blade stall and one other undesirable effect, flapping (described in Chapter 6).

Look Out for Zero G's and Porpoising

There are two conditions that can occur in a gyroplane which can be deadly. They are both avoidable, as you will see:

4. Zero G's - Have you ever ridden in a roller coaster and felt yourself floating off the seat as you went over the top of the hill? That's "Zero G's"—a condition where your weight is temporarily zero. It's great fun in a roller coaster, but it can be deadly in a gyroplane.

Remedy: Here are the words of an authority on gyroplanes, Dr. Igor Bensen: "A gyroplane must never be flown in zero-G condition for more than two or three seconds at a time, because rotor speed decays when no air goes through it. Short duration zero-G and negative-G excursions are not dangerous and occur often in gusty weather without any ill effects. So zero G's and negative G's are not something to be feared, but they should not be deliberately induced by the pilot for longer than two or three seconds."

You should avoid flying at zero G's, even for an instant. Fortunately, a zero-G condition is not difficult to avoid. Don't try aerobatics or extreme maneuvers in a

gyroplane. Don't do "over the hill" maneuvers (sharp
roller coaster-like movements in the air). Don't do sharp
pull-ups or violent turns. Keep your maneuvers simple.
Just flying a gyro is thrilling enough!

5. PIO - Pilot induced oscillation or PIO will concern you
when you begin to make flights around the pattern and stay
in the air for longer periods of time. PIO (also called
porpoising) describes an unplanned roller coaster-like movement in
the air,
with the
machine
alter-
nately
climbing
and
diving and the severity increasing with each dip. It can be
disastrous when not controlled, a problem which has led to a
great deal of criticism of the gyroplane.

Is the gyroplane at fault, as some people say? Igor
Bensen has said 'No,' that it's the inexperienced pilot who
induces porpoising, not the machine. It's certainly a greater
problem for the inexperienced pilot who hasn't yet learned
the unique handling characteristics of the gyroplane. The
pilot's reflexes must be developed to handle the rotorcraft
smoothly without thinking about it.

What happens in PIO is that the untrained pilot is too
slow on the control stick. As the machine noses up, he
corrects by reducing power and applying FORWARD stick.
But the machine, acting like a pendulum hanging from the
rotors, already is about to swing down. The pilot's actions
merely make the resulting dive worse. He corrects by adding
power and giving BACK stick, but again he does this as the
machine has already begun to swing back up. So the resulting climb gets steeper and he continues to make the climbs
and dives worse and worse.

My experience with PIO, both intentionally-induced and
accidental, indicates that it can be overcome by the pilot.

80

However, an improperly-trimmed rotor head or loose pivots in the control system can aggravate the problem, requiring greater skill to recover.

Remedy: Have enough dual instruction time to develop good reflexes on the stick and to learn the quick rhythm of the gyro. If you learn on a tow line, remember that the tow line itself damps PIO, so later you will still need plenty of low-level practice when you add engine power.

To get your rotor head properly trimmed, adjust it until you can maintain your proper cruising airspeed (50 mph for a typical gyroplane) with a neutral stick feel, while staying just above the runway. Then on your early trips around the field, maintain your cruising airspeed exactly, holding the stick very lightly with two or three fingers. Make a conscious effort not to hold it tightly like a sledge hammer. Remember, "white knuckles" means tension, and tension can lead to porpoising.

If, despite your best efforts, you find yourself in PIO, reduce the power. Once the PIO stops, if you have enough altitude, add power to resume cruising speed and continue on. If you're at a low altitude and behind the power curve when you finally stop, add power and make the best landing you can.

PIO, like any other oscillation, needs power to be sustained. Remove the power and it will die out. Keep your eye on the airspeed indicator and strive to maintain a constant speed whether climbing, flying level, turning or descending.

Chapter 10

Safety Checks

"Switch on!" You've already primed the engine and pulled the propeller through a couple of turns. You reach toward the engine more cautiously this time, knowing this pull may set the prop singing if the hot spark does its job of releasing the explosive fury inside those cylinders. You feel your stomach tighten and your inner ears harden against the noise they are about to hear.

This is the moment you've been anticipating. You've built your machine carefully, spent hours learning to fly a dual-seat machine, and today you've brought your machine to the flying field for your first solo powered flight.

But wait! Are you sure your machine is ready for that next monumental turn of the prop? Once it fires you move into a different world from the one you were in while building and testing your machine. The building is over. It's too late to correct an error when you're in the air under power. Now you must concentrate on flying, confident that your machine is 100% ready. If it's not, "Switch off!" Don't turn over that engine—yet.

To get that 100% confidence, there's no shortcut for certain safety checks that are recommended by designers. Here's a list:

Hang test - This is one way to make your first flights easier and safer. If your machine is nose heavy or tail heavy, you'll have trouble getting proper airspeed. You may be taxiing like mad, never quite getting off the ground. Or you may horse it into the air behind the power curve and discover that flying that way requires considerably more skill than anyone has on his first solo powered flights. And if your machine is badly out of balance, it may not be safely flyable at all.

Besides making sure your first flights are not made under a handicap, a hang test is required for your Weight and Balance Statement if you license your machine. It's not hard to string up your gyro by the teeter bolt under a tree or a rafter with a degree-reading level taped to the mast, having someone read it while you sit in flying position. The angle indicated by the level should be within standards set by the designer.

Rotor blade pitch - Because that engine will be singing so hard for your attention, it's easier to ignore the quieter demands of your rotor blades when you move into powered

The blades must be assembled carefully.

flight. Naturally your rotors must be free of structural damage, nicks, cracks or dents. But they must also have proper pitch. This can be checked by mounting them on the hub, turning the whole blade assembly over (to approximate the effects of gravity and the inertia of spinning), clamping it to a solid level surface, and setting a degree-reading level at one-foot intervals along each blade. This will tell you whether the blades are identical.

Spanwise balance - While you've got the blade assembly hooked together, turn it back right side up. Suspend it from its teeter bolt and make sure your blades have proper spanwise balance. If one blade drops, even slightly, it must be corrected. A penny dropped on either blade tip should make it go down. Follow the instructions of the blade designer to balance them exactly.

Blades in pattern - Your rotor blades should be exactly opposite each other in a straight line when viewed from the top. With the blades fastened to the hub, run a string from the top of the leading edge of one blade to the same spot on the other blade. The string must pass exactly over the center point of the hub.

Chordwise balance - Most people don't check manufactured blades for balance front to back, even though this is good protection against rotor blade stall. Sometimes designers don't publish the chordwise balance point of their blades. Except for homebuilt blades or blades of questionable origin, most gyro pilots rely on manufacturers to build blades that are properly balanced chordwise.

Rotor head condition - Since you've probably been doing some taxiing around in your gyro, it's time to check carefully for good rotor head condition. There should not be any "slop" or looseness between the parts.

Control stick tightness - With a joystick, you have a whole series of connections to check for tightness. If you have an overhead stick, check for tightness where the stick bolts to the head. Taxiing on grass is especially hard on control stick connections, and they should not have any noticeable play or looseness.

Rotor hub creasemarks - Look underneath for rotor hub creasemarks due to flapping against the stops. Since any crease is a potential fatigue point, carefully polish out any severe marks (or get a new hub if it's really bad).

Rotor head trim - Get this as near to a proper setting as possible. Designers should be able to suggest how to adjust the trim for your first flights. An out-of-trim rotor head poses problems similar to an out-of-balance machine, so it definitely needs to be as right as possible for the beginning pilot.

Engine checks - Grab the prop by the hub and shake it. The engine mount should be solid, with no play whatsoever (You may notice springy movement due to the resiliency of your metal frame). Check your throttle cable to see that it works freely. Look closely at the engine end of the cable. The spring must pull the cable back when you close the throttle. The cable must not push the throttle closed. The cable is tremendously strong when pulling, but pushing is not its favorite thing. Ask it to keep pushing and it may stop working while you are in the air. Also, make sure your magneto switch works. It's your only method for stopping the engine immediately. You may want to add another kill switch on the instrument panel or near the throttle control. Be sure to mount the kill switch where you are not likely to turn it off accidentally.

A careful preflight inspection is a must.

These engine checks cover some of the most troublesome problems. There are other ways to assure you have a healthy engine, like a static thrust check. As a good rule of thumb, your static thrust should be equal to at least one half your gyro's gross weight. You need to have confidence that your engine is producing enough power and that it will give you that power every time you call on it.

Checklist - On the next page is a suggested flightline checklist for your pre-flight inspection. It was developed through practical experience by the author.

Pre-flight Checklist

❏ Pockets .... Empty

❏ Gas tank .... Full

❏ Airspeed indicator . Uncovered

❏ Drift indicator .... Mounted and operational

❏ Rudder pedals .... Proper movement

❏ Front wheel........... Centers itself
Tire inflated

❏ Rudder cables........ Free, not frayed

❏ Main wheels .... Tires inflated
Normal play on axle

❏ Seat belt............... Open and ready

❏ Throttle................ Movement free and easy

❏ Rotor head .... No looseness
Complete movement in all
directions
Stick mounting secure

❏ Engine mount........ Solid

❏ Engine covers........ Removed

❏ Propeller .... No nicks or cracks
Washers snug

❏ Rudder................. Hinge is sound
Free movement

❏ Rotors .... No nicks, cracks or damage
Proper rudder clearance

❏ Rotor mounting ..... Mounting bolts and teeter
bolt secure

Chapter 11

Getting Off the Ground

One of the problems with flying a gyroplane is that you have to do the hardest part first: the takeoff. And right after the takeoff lurks the second hardest part: maintaining level flight. But at the end of each flight comes the part that's easiest of all and downright fun when you know how: the landing.

"What!" I hear a fixed-wing pilot saying. "That doesn't make sense. Landings are the hardest part of fixed-wing flying! And who ever heard maintaining level flight was difficult?"

This is one example of the peculiarities of gyroplanes. In these machines, both the takeoff and level flight are difficult enough to require lots of practice before they can be done safely solo. The landing requires practice, too, but dual-seat training is good preparation.

This chapter will be especially helpful to the person who's learning to fly on a towline gyroglider. It suggests a step-by-step procedure for learning to take off in your own single-seat machine, after you've mastered the towed gyro. For those who are learning on a powered dual-seat trainer, this chapter will also serve to give you a procedure for getting into the air in your own single-seater.

Make sure you have an expert observer watching you when you first fly solo. If possible, have your instructor there when you make your first solo flights in your own machine. Instructors are usually happy to supervise your first solo flights, since they know they can help you succeed more safely.

Regardless of the type of dual-seat training you're having, you should proceed cautiously in learning to fly your own single seat gyro. It's the riskiest part of your gyro flying experience, those first hours when you sit alone on your machine with no one else aboard to help you.

With a well planned step-by-step approach, you can be successful in getting into the air in your own gyroplane. I've done it the hard way, with my training done on a towline gyroglider and my first powered flights made alone in my own machine, watched by an experienced observer. I paid a lot of attention to the advice and experience of several people who had successfully learned to fly a gyro. The advice in this book is also based on the sadder-but-wiser experiences of several individuals who did not follow a step-by-step approach and paid for it in mishaps, bruises and broken rotors. Worse yet, it only took one or two mishaps to discourage these individuals from going on, and they never knew the thrill they had set out to experience—flight in a gyroplane.

Steps to solo flight

Taxiing without blades - You probably don't need this step if you had a complete training program in a dual-seat powered gyro. It used to be a required step for towline glider students getting used to their powered single-seaters. The idea is to leave the rotor blades off and taxi around for a while, never exceeding 15 miles per hour. This starts to develop the reflexes you'll need to fly. Operating the throttle, for example, may be different on your own machine from the powered trainer. You need to know without thinking which way to move the throttle for more power, which for less.

You need to be used to steering your nose wheel. Taxiing around will begin building proper reflexes. If your rudder pedals aren't linked to the nose wheel, you can work on the proper reflexes by taxiing on a smooth hard surface, steering with your feet on the rudder pedals and giving small bursts of throttle when you want to change direction.

One caution: Keep your speed down with the rotors off. On the ground the gyro is normally stabilized by its whirling rotors. Without these rotors you're riding a big overpowered tricycle. You must slow down to a crawl before turning sharp corners, and your speed should never get above about 15 miles per hour.

You can move on to the next step when you can speed up or slow down the engine without thinking which way to move the throttle and when you can instinctively push the proper rudder pedal for the direction you have in mind.

Taxiing with blades - Now comes your first experience with your complete powered machine! With your rotor blades mounted and spinning, you will face much greater wind resistance than if you taxied without blades, requiring more throttle. The old skittish tricycle feel is gone, since the spinning rotor seems to pull backward like a big drag chute.

Begin by doing everything straight down the field or runway. With your engine running, face into the wind and spin up your rotor blades. Hold full BACK stick (rotors tilted fully to the rear) and taxi straight ahead at about 10 MPH airspeed. Try to find the slowest speed that will maintain a modest rotor speed. This will show you the speed you will use in taxiing back for your next run.

Next time, increase your speed to 15 mph. Little by little your rotors should begin to pick up speed, sounding more like they did when you flew in the dual seat trainer. If your rudder pedals aren't linked to the nose wheel, shift your feet to the rudder pedals as soon as they provide effective steering, jazzing the throttle occasionally to help get effective airflow over the rudder.

It's extremely important from now on that you watch for flapping, the bumping feel in the stick that means your rotors aren't yet ready for so much forward speed. Flapping is less noticeable now that you have engine noise and vibration to compete for your attention, but it's just as important as ever. If you encounter flapping, slow your forward speed or reduce BACK stick, or both. The trick is to increase your forward speed *gradually*, never more than your rotors are ready for

(Here's where your training experience pays off again!).

Make more runs at gradually increasing airspeeds. When you reach about 20 MPH or so, the machine will suddenly sit back on its tail. This is your first sign that you're generating lift. It only took about 60 or 70 pounds of lift to raise the nose wheel, so you're not likely to fly just yet. Reduce power enough to drop back onto the nose wheel. Do this same sequence of rocking back on the tail wheel several times so that you get the feel of what it's like when the nose gets light just before it rises.

On your next run, try to anticipate when the nose is getting light. This time, instead of continuing to hold full BACK stick, move the stick a bit more toward the neutral position so that the nose stays lightly on the ground, but your forward speed increases slightly. By gradually reducing the amount of BACK stick and increasing power as necessary, work up to about 30 MPH, keeping the nose wheel lightly on the ground.

As you approach the end of the field, first reduce power to an idle and gradually tilt the rotor fully back (full BACK stick) to work as a brake (Don't throw the rotors back suddenly, as this could pop you into the air). Then move your feet to the wheel brake as needed. If there are weeds or runway lights near you as you turn around, tilt your rotors flat to the ground (full FORWARD stick) to clear them. Then as you taxi back, resume full BACK stick to keep the blades spinning. Do this sequence over and over until you feel fully in control of the situation. Then quit for the day. Whether you've done all your solo practicing on one day or spread it over several days, this is a good time to give yourself an overnight rest. One secret to safety is to master one step at a time, and be well rested when you move on to the next.

Here's some advice from the man who taught the world to fly Gyrocopters, Dr. Igor Bensen: "Statistics show that more accidents happen to impatient pilots who try to cram their learning cycle into too short a time. It just can't be done. Your body and mind can absorb so much new information at a time. You must make up your mind, and stick to it, to practice no more than one hour in the morning and one hour after lunch. Then stop. Any rushing of this learning program can lead to a disaster, or at least a broken rotor."

Breaking ground - The next time out, you will continue the previous exercise, gradually increasing your speed until you fly. From now on you must make a point to keep the stick centered left-to-right regardless of how you move it up or down, so that when you lift off you will fly straight ahead.

If there is a slight crosswind, learn to apply a steady pressure on one rudder pedal and, once airborne, slightly tilt the stick to keep the gyro from drifting sideways. (It's best to avoid a crosswind at this time, if you can, to simplify things.)

For your first experience in leaving the ground, try a speed of about 35 mph. Build up to this speed, keeping the nose light. Your rotors should sound and look about as they did during your takeoffs in the dual-seat trainer. The wind in your face and past your helmet will seem fairly strong. As you reach about 35 mph, begin to increase the amount of BACK stick. You probably won't notice the exact moment you leave the ground, so think about your control movements: Keep the nose straight ahead with the rudder pedals. Keep the machine level with gentle side-to-side movements of the stick. Make throttle changes gradually.

As soon as you discover that you are in the air, keep controlling the machine and gently shave back the power to an idle. Flare with a gradual BACK movement of the stick just as you did in the dual-seat trainer and land touching the tail first. You should not have gone higher than six inches or a foot on your first try, but it will seem much higher!

If any control can be emphasized at this point, it's the rudder. This is especially true if you learned on a towline trainer, where the machine is steered straight ahead automatically (so long as you are directly behind the tow car). Now it's entirely up to you to keep the nose pointed in the direction you're actually going. As you know, the gyro can't land in a crab. Now it's your job to see that you're aimed right when you make contact with the ground.

The altitude of your first gyro flight will probably be measured in inches and the distance will probably be about 25 or 50 feet. At 35 mph or so you will take off in a somewhat nose-high, flared attitude, ready for a landing. By reducing power as you break ground and by tilting your rotors back, you simply continue to flare to a landing.

The next step is to increase speed and transition to a cruise condition in the air, then back into a flared landing condition. We can look at two methods to accomplish this: 1) the traditional takeoff and 2) the "flyaway" takeoff.

The Traditional Takeoff - This is the method that was used originally, when the gyro first became available. This method uses engine power to transition from a slow-speed takeoff to cruising flight. The takeoff speeds are as low as the speed you just used to break ground. But at takeoff, instead of reducing power, a high throttle setting is maintained to keep the machine off the ground and gaining speed in the air. At 3 to 5 feet altitude the nose is brought down for level flight, and as cruising airspeed is reached (about 50 mph for most gyros) the throttle is reduced to a cruise setting. A highly experienced pilot with a good machine can use this method to take off at less than 30 mph airspeed, making a very interesting demonstration as he "pops" quickly off the ground.

The Flyaway Takeoff - This method is more like the way powered trainers take off, and is basically the way airplanes get off the ground. It uses airspeed instead of engine power. Beginning pilots sometimes use this after they have had trouble mastering the traditional takeoff. To accomplish the "flyaway" method, you take off by increasing airspeed while rolling along the ground, keeping the nose light until about 45 MPH is attained. At about that speed the machine will "fly away," breaking ground gently and gradually in very nearly a cruise configuration. Very little lowering of the nose or throttle reduction is needed after takeoff to maintain level flight.

Which method should I use for takeoff? Either method works. One or the other of these will seem more like the method used by your instructor in the dual-seat trainer. Powered trainers usually use the flyaway method, while towline trainers usually use the traditional style. If you try the traditional takeoff but have difficulty, you may find that a switch to the flyaway method will help. Flyaway is a good way to break a habit we develop on the towline: taking off at low speed and flying nose high. This appears to be a common problem among people who learn in a towline gyroglider. With the exciting rush of wind, the swish of the rotors and the roar of the engine, there's a strong tendency to cling to a low-speed, nose-high flying attitude. It just feels safer somehow. Unfortunately, it's not. Flying nose high requires exceptional quickness and skill on the rudder, since the machine wants to tilt to one side and is unstable with respect to speed (This is an example of flying on the back side of the power curve).

For a beginning pilot it's often easier to make most of the major adjustments of throttle and stick on the ground, rather than in the air. When you "fly away," your throttle is already at or near a cruise setting, your nose is already down near a cruise attitude, you've already gotten your rotors to flying speed, and you've already gotten a feel for how loud the wind and engine noises are near cruising speed.

An advocate of the traditional takeoff would mention that method's lower takeoff speed and shorter takeoff run, and would point out that the greater control of throttle and airspeed which are required are skills that beginning gyro pilots must learn anyway. With the conventional method, if the pilot does fly nose high, he already has a good start on his flare for a landing. As long as he's near the ground he can react to trouble by reducing power and landing immediately.

Whichever method you use, spend a lot of time, at least three hours, practicing your takeoff, flying level at 3 to 5 feet altitude straight down the field, and landing (Remember, the takeoff is one of the most difficult parts of gyroplane flying). Any time you feel you may be in trouble, simply shave back the power and flare to a landing (All throttle movements must be gradual). Now you're well on your way, so give yourself the rest of the day off to celebrate!

Chapter 12

Trouble Shooting

When the United States started trying to put astronauts into space, the spacecraft designers applied all the laws of physics, along with a new one that seemed just right for the occasion. This new law of physics didn't have a fancy name like the old ones. While every engineer had heard the name of Isaac Newton, who formulated many of the physical laws they had memorized in school, nobody knew exactly where this new law had come from. So, not only did they invent a new law of physics, they also invented its author, whom they named "Murphy."

They discovered that "Murphy's Law" was very useful in designing and building spacecraft. You'll find that this same law is very useful to gyronauts, too. Here's what Murphy's Law says:

> *Anything that can go wrong*
>
> *will go wrong.*
>
> *—Murphy*

In your early attempts at flying, you'll probably experience Murphy's Law. You can bet that something will go wrong, and it may seem like everything that could go wrong already has. But you haven't heard the last from Murphy. According to his law, you're going to continue to have problems to solve.

During your training in a powered two-seater, your instructor may have covered all the possibilities we're talking about here. But when you're on your own in the single seat of your own machine, with no one else aboard to help you analyze what's happening, you'll probably run into problems you'll have to figure out.

If it's any consolation, this isn't unusual. It seems that almost everyone has some special problem in learning to fly a gyroplane. And now that there are thousands of us out there going through the same experiences you're now having, you don't need to feel that your problems are different. In fact, you'll have a hard time coming up with a problem that somebody hasn't already faced and solved.

So, instead of taking the lonely path to finding a solution on your own, we're going to list some of the most common problems people like you have discovered. And we'll suggest some solutions. This should help you get past the glitches and move on to the real fun of trouble-free flying.

But keep in mind that no book can cover everything, and there's no way for a book to know exactly what your problem is. So the best solution to problems is to get an experienced gyro pilot to watch your early flying and give you a report on what he sees. Often someone else can identify a problem that you yourself can't figure out. And a good, experienced gyro pilot can also suggest some good ways to make things better.

Note that I specified a *gyroplane* pilot. You should only trust the advice of someone who has had a lot of experience actually flying gyros. No one who has flown only airplanes—or only helicopters—can give you correct advice. It doesn't matter whether he's a flight instructor, an airline pilot or an astronaut. If his experience isn't in gyroplanes, he'll probably tell you wrong, no matter how good his intentions are.

And make sure your observer is an *experienced* gyro pilot. I've heard some pretty authoritative-sounding talk from people who "used to fly a gyro" or who "once built my own gyro." Have you seen him fly? Is he good—I mean really *good*? If so, then he's your man!

Having an observer is important even after you've learned to fly in a powered dual-seat machine. It's the way to make sure your first flights in your own single-seat machine go well. If you're learning by the towline gyroglider route, then getting a critique is a necessity. Murphy says you *will* make mistakes. So take Murphy's advice and anticipate them.

Starting on the next page is a list of common problems you can anticipate. You'll notice that most of these happen on takeoff. That's because getting off the ground is the hardest part of gyro flying, and the most likely place you'll find difficulty.

Read these over so that you understand them. Then, when your expert observer points one of these out, you'll already know what he's talking about.

Common Problems

1. "I just can't get off the ground."

Airspeed may be too low - Anything less than 30 or 35 MPH may not be fast enough for a takeoff. It takes airspeed to get your rotors up to flying RPM's and to lift off. Anything above 50 MPH is probably too fast and may indicate a problem with the aircraft rather than with your flying technique. Consider some of the checks in Chapter 10.

Rotor speed may be too low - Your rotor blades must reach a certain speed before they generate enough lift for takeoff. Prerotators don't turn the blades fast enough for liftoff; you still have to taxi a bit to get your rotor speed high enough. Also, if there's a delay between your spinup with the prerotator and your attempted takeoff, you may have lost some rotor speed, which has to be made up in taxiing.

If you fly without a prerotator, you'll have to do lots of taxiing at gradually increasing speeds to get your rotor blades turning fast enough to fly, especially on days when there's not much wind. If you have very gradually taxied up to 45 mph without being able to lift off, there may be problems with your rotors. They may have too much pitch to get started easily, or they may not be properly designed and adjusted. Try the rotor blade checks in Chapter 10.

Not enough power - It may take a lot of throttle to get your rotor speed and airspeed up, especially with lower-powered engines. People who learn on powered trainers have an advantage in knowing about how much throttle it takes. For most engines, you may have to go to full throttle, but do it gradually, watching your airspeed and watching for rotor blade flapping. If you can't work up to flying speed, a static thrust check may be needed to determine whether your engine is sick or your propeller is not right.

Rotor head may be out of trim - The popular offset gimbal rotor head has tension springs which must be adjusted for proper stick feel. If the springs are missing or the tension is grossly wrong, you may not be applying as much BACK stick as you think you are, or too much. Check the designer's recommendation for a trim setting.

Machine may be out of balance - If your machine passed the hang test, this should not be a problem (unless you've changed something since the hang test). If it didn't pass the hang test, shame on you!

High density altitude - This is another way of saying "thin air." It can result from any of these three causes: 1) a high elevation, 2) a hot temperature, and 3) high humidity. If you have any of these conditions, you won't get as much thrust or lift. It usually takes all three of these to keep you from lifting off, and this doesn't happen very often. If you can't change the conditions, you might try a change in rotor blades for more lift, such as using a higher pitch or longer blades.

Machine may be too heavy - Bathroom scales under the wheels will show whether or not your machine exceeds the designer's maximum gross weight figure. Don't forget to figure in the weight of the rotor blades, the rotor hub, a full tank of gas and you, fully clothed. You may be able to increase your allowable gross weight by increasing the pitch of your rotor blades. If your machine is designed for longer blades, you can extend the rotor hub or use longer blades.

2. "I got off the ground, but couldn't stay up."

This could be due to any of the problems in the "I just can't get off the ground" paragraphs just above. One of the most common errors is flying nose high, never getting the airspeed up to cruising speed. This is especially likely for people who started in a towed gyroglider. Some symptoms of this error are: having to "horse" the machine off the ground at low speed with a vigorous BACK stick motion or high rudder sensitivity while in the air (The machine seems to try to yaw one way or other due to the nose-high slow-flight attitude.). Re-read the paragraphs on Behind the Power Curve in Chapter 9.

3. "I left the ground, but had trouble controlling it."

Gusts - The best way to deal with gusts at the early stages of powered practice is to avoid them. Try for calm air for your initial practice. You need to be able to cope with gusts even tually, but avoid them until you feel confident on the basics of takeoff, level flight and landing. Flyers of the Wright brothers' era used to get up early in the morning to take advantage of the calmer, denser air. It's not a bad idea for you, too.

Rotor head may be out of trim - This was covered a few paragraphs back, except for lateral trim. Since many rotor blades have a slight tendency to roll toward the retreating blade side (In the United States this is normally toward the left), the tension springs on offset gimbal rotor heads are sometimes moved to the right to neutralize this effect. If this adjustment isn't properly made it may produce light rolling tendencies, especially when maneuvering or riding through gusts. This makes control more difficult.

Not used to flying without a tow rope - If you learned to fly in a towline gyroglider, you may not have your reflexes trained well enough to do without the tow rope. It's difficult at first, suddenly having to maintain your forward heading yourself with the rudder. The solution is either to get training in a powered dual-seat machine, or spend more time practicing taxiing with rudder control.

Machine may be out of balance or too heavy - Either condition makes control difficult or impossible, depending on how bad it is. Read the paragraphs just before this under "I just can't get off the ground".

4. Ballooning

You're rolling along for takeoff. The machine doesn't seem to want to break ground soon enough, so you give it a good dose of BACK stick. For an instant it seems like nothing is happening. Then suddenly the machine noses up and leaps into the air, much higher than you had planned. That's what "ballooning" means in this case. To prevent ballooning, make your acceleration gradual. You should be able to ease off the ground with very gentle BACK stick movements, and very little excitement. Consider the "Flyaway" takeoff method explained in Chapter 11.

5. Other problems

Every month, all over the country creative gyroplane pilots are inventing new problems for themselves. It would be impossible to keep up with all these new innovations in this book. In fact, this chapter does not pretend to solve problems, but tries to suggest some ways that may possibly be effective in dealing with them.

The actual solution is up to you, you "experimental pilot". But take heart, Murphy's Law didn't keep the astronauts down, and it needn't keep you down, either!

Chapter 13

Keeping it in the Air

When you make your takeoff and get away from the ground, things change. Your gyro stops being a three-wheeled ground vehicle and becomes a flying machine. That brings changes in the forces acting on your vehicle and requires you to learn a new set of rules.

Flying straight and level introduces the importance of airspeed control. It's different from your takeoff, where the airspeed is constantly changing. In straight and level flight your airspeed should stay exactly the same.

If you're flying in your own machine, having hopped off the ground in a successful takeoff, take a look at your airspeed indicator. It ought to show the proper cruising speed, which is about 50 mph for most of the popular gyros. If it's less, you must get the nose down and increase power so that your airspeed comes up. If the problem is persistent, re-read the explanation of "Getting behind the power curve" in Chapter 9. Or try the Flyaway takeoff method in Chapter 11.

Let's assume you have everything stabilized in straight and level flight. (If you've learned on a towed

gyroglider, your altitude should be only a few feet at this point, and your flights should be short hops, landing before the end of your takeoff strip.) Now let's see what happens when you climb or descend.

Suppose you make your takeoff and level out a few feet above the ground. Now climb up to 10 or 15 feet, then fly back down to about a foot above the ground, shave back power and land. You will have discovered that to go up you increase power. To come down you reduce power. This is a very important discovery: **POWER CONTROLS ALTITUDE.**

If you tried to climb merely by giving it BACK stick, you would find that this just raised the nose and slowed you down. If you drop the nose, that speeds you up. So you've made another important discovery: **THE STICK CONTROLS AIRSPEED**. How about that!

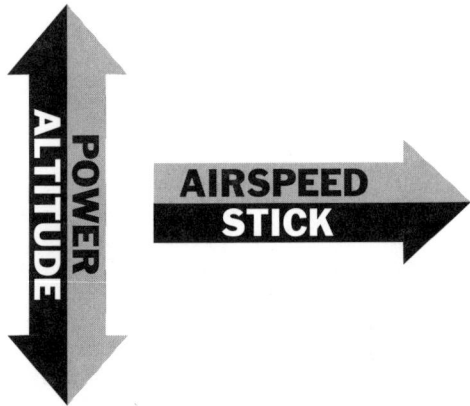

And, while you're discovering the basics, here's another one: **LANDINGS START WITHIN A FEW FEET OF THE GROUND**. To land, you shave back power slightly (never chopping it, since that makes the machine lurch and complicated things) and fly down to within a few feet of the ground, *maintaining nearly full cruising airspeed*. Then you gradually reduce power to an idle as you flare to a tail-first landing.

You may make another observation if you fly when the wind isn't blowing directly toward you. In a crosswind the machine seems to want to fly somewhat sideways above the ground. If you use rudder to make it head straight down the field, you must bank slightly toward the wind to keep from drifting to the side of your intended flight path. The answer is to go ahead and let it fly sideways (or "crab"), with the nose angled toward the source of the wind. That way the machine is actually flying straight and level through the air. But since the air is moving sideways across the field, you actually follow your intended flight path straight down the field. You can check yourself by glancing at your drift indicator (Most gyros have a small flag or a piece of string out front to let you know if you are tracking straight.). If it points back at you, you're flying straight through the air regardless of how it looks compared to the ground. If the drift indicator doesn't indicate you're flying straight, use the rudder to correct it. Of course, when you land the tailwheel should touch first and the machine must be straightened out to point toward its path over the ground, since *the gyro cannot land in a crab.*

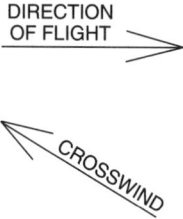

DIRECTION OF FLIGHT

CROSSWIND

Now it's time for some gentle "S" turns, banking first left, then right. If you're learning on your own, always return to your straight path down the field and descend to land. These turns are made with an ever-so-gentle side pressure on the control stick. Your nose will come around and point toward whatever direction you're turning, probably without any help from you on the rudder pedals. This is the "weathercock" tendency, with the airframe acting like a weathercock and swinging around to point in the direction of the relative wind.

As you increase your angle of bank you'll discover that to maintain altitude in a steeper turn you must apply more BACK pressure on the stick as you turn, and slightly more power. In a bank at a 20 degree angle this becomes quite noticeable. Get this connection firmly in mind, since at higher altitudes it's harder to notice. Up there, you've got to remember to **ADD BACK PRESSURE ON THE STICK AND ADD POWER DURING TURNS**.

In your own machine you must be able to climb, descend, and make well-banked S-turns comfortably. If your field is long enough, increase your altitude to 50 feet or so before coming back down to land before the end of the strip. If it's shorter, go up as high as you can while getting back down comfortably. But until you and your own single-seat machine are ready, don't make the mistake of getting too high to land at the end of the field. Every so often this happens to a beginning gyro pilot. If he's lucky he may complete his first circuit of the field. More typically he will end up with a machine (and his confidence) to rebuild. Your first circuit should be when you feel you're ready, not before.

Before your first solo circuit, have you learned how to cope with gusts? In a powered trainer your instructor was there to help you if things got rough. In your own single-seater at low levels you could react to a gust by reducing power and landing if you were over the landing strip. But on your first solo circuit you must plan to fly right through gusts, and you must have the experience and confidence that you can do it. It's best to make your first solo flights at a time when the air is calm, as in early morning or late evening. But even then you must feel confident you can stay in the air all the way around.

At last you feel the time is right and you and your single-seat machine are ready. On your first pass down the field, if you can find some rough terrain on which you would definitely not like to land, like some tall weeds or a crop, venture over that terrain and then back to your normal landing spot, without going around. How did it feel being all by yourself over "unfriendly" territory? You're going to have that feeling several times as you go around the pattern, so make enough trips above the rough stuff to get used to it. Naturally, you're better off making your first trips around the pattern someplace where there will be a maximum amount of friendly terrain, but not everyone has a dry lake bed to use for his flying field!

What's it like to make your first solo flight around the field? Let's follow a typical experience and see.

105

Your First Solo Flight Around the Field

You feel that you and your gyro are ready. You're confident. You fire up the engine, strap in, spin the blades, and
taxi to the end of the field. You line up into the wind and
prerotate your blades to a good speed. You pause just a
moment to take a deep breath. Then you crack the throttle,
picking up speed. The blades are swishing. Airspeed 25, 30,
35. Blades singing, engine throbbing. Lift off!

You climb to 10 feet. Everything is okay, climbing on up
to 30, 40, 50 feet. Level out, flying straight and level, speed
exactly 50. Looking straight ahead at the horizon. Approaching the end of the field, you're committed now.

Ease into that first turn. Shallow bank, nudge BACK
stick, a fraction more power. Banking steadily, turning,
watching the airspeed. Got to keep it at 50 no matter what.

Ready to turn downwind, parallel to the takeoff roll.
Concentrating on the horizon, the airspeed. Still going 50.
Forget groundspeed. Got to keep that airspeed showing 50.

Now level, flying downwind. *Halfway around!* Whump!
There's a gust! Just felt like a bump in a car. Flew right
through it!

Still heading downwind, now even with the landing
point. Ease into a bank toward that point. Still holding 50
MPH. Coming on around toward the point.

Approaching the point with the landing spot in mind.
Shave back the power now, lower the nose to descend. Still
holding 50. Down to 15 feet, 10 feet, here it comes!

Down to a foot. Level out. Crank back power and flare.
Less power, steeper flare. Less power, steeper. Engine idling.
Touch the tailwheel, steer straight. You're down! You made it!
You've flown the circuit!

It will be one of the great accomplishments of your life,
that day when you make your first solo flight around the
field. The feeling can't be described. But it will let you know
that all the practice, the step-by-step work and study paid
off. It was worth it. You did it yourself. You flew! Tell the
world, you flew!

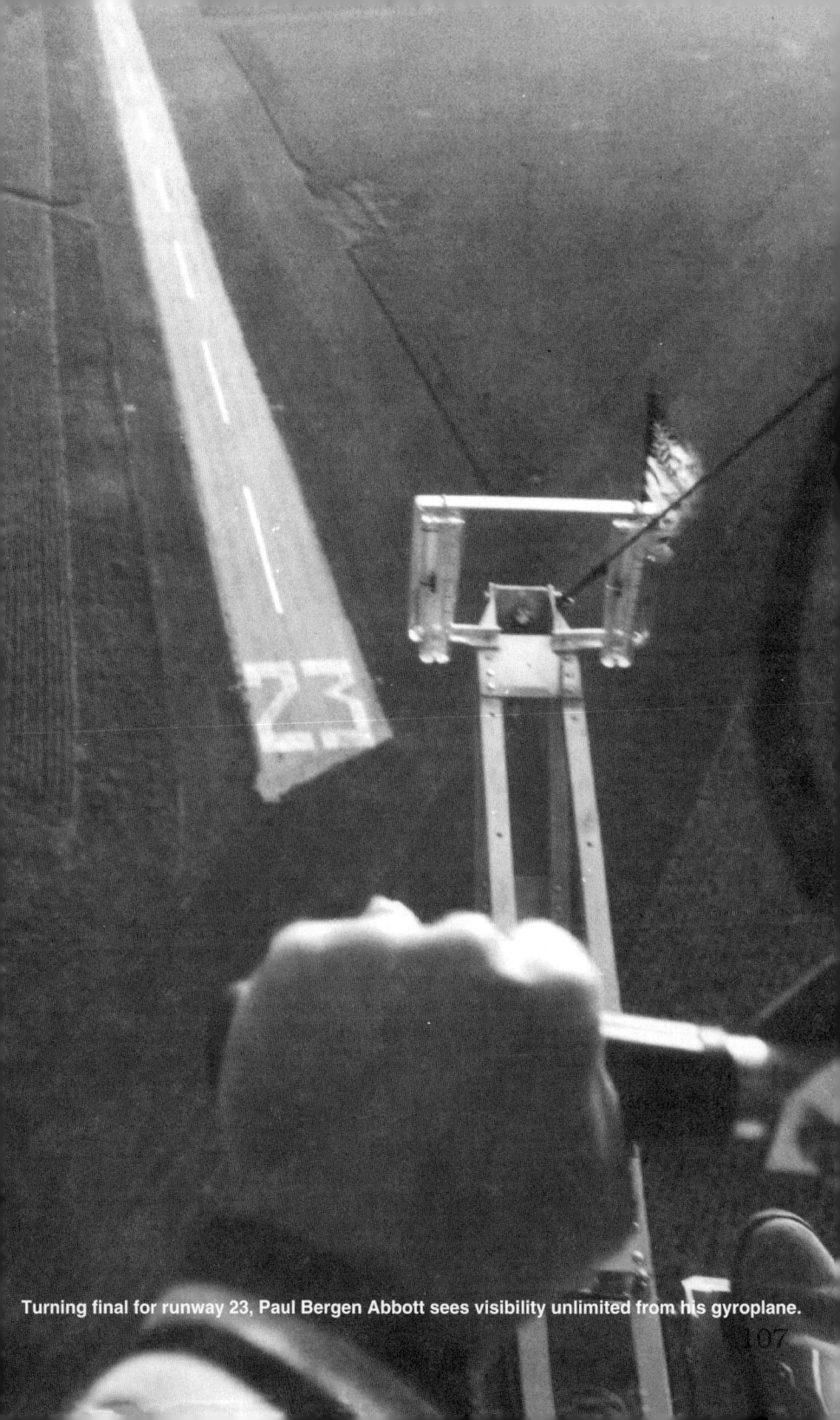

Turning final for runway 23, Paul Bergen Abbott sees visibility unlimited from his gyroplane.

Chapter 14

How Safe is a Gyroplane?

You've heard the glowing reports: gyroplanes are thrilling to fly, sports cars get you there in style, new medicines will relieve your ailments.

But it seems that everything that gives you great benefits also has some serious drawbacks. A sports car can enhance your image, attract pretty girls, outperform other cars—and wrap you around a telephone pole. Wonder drugs can relieve pain, stop infections, cure diseases—and kill or disable you when misused. The gyro? Let's stop and consider that one a little more closely.

Everybody likes to talk about the positive side of gyro flying. That's what most of the rest of this book is about. But to help you prepare realistically for this new experience, we need to take time out to do some negative thinking. There really is a detrimental side of this sport, and you should give it some serious thought as you learn about it.

This is not meant to discourage you. In fact, the more realistic your knowledge is, the less likely you are to be discouraged and the more likely you are to succeed.

First, let's state a simple fact: There's danger in many things we do. We sacrifice more than 40,000 lives each year in the United States in our automobiles. Good statistics on gyroplane accidents aren't available, but they would undoubtedly be much lower.

You can get hurt flying a gyroplane. You can be killed. That's the grim truth, told to us in tragic stories of people who have flown before us, who fly no more. In the solemn quiet of a funeral service for a fallen friend, you think deeply about whether to fly again. I've had that experience, and came away convinced that my place was back in the air under whirling rotors.

Many gyro accidents go unreported, making it difficult to assess today's true accident rate. However, we can get a clue from statistics collected before ultralights existed, which show that homebuilt gyroplanes of the Bensen Gyrocopter variety were safer than factory-built airplanes. Official government statistics showed that airplanes had 5.6 accidents per 100 aircraft per year, while Gyrocopters recorded only 3.6 accidents per 100 per year.

How crashworthy is a gyroplane?

When you consider that a gyroplane can hurl you through the air like a bullet and elevate you to heights known only to the angels, it's obvious that this vehicle has the capability of wiping you out. So can a car. So can an airplane. So can a bicycle.

In some ways a gyroplane isn't as likely to crash as other aircraft. In case of engine failure, you simply glide to a landing while the rotors keep turning. This is different from a helicopter, where an engine failure means an urgent de-clutching, a rotor blade pitch change and a very critical transition from powered flight to autorotation.

The gyro lands very slowly, sometimes at zero airspeed, and in very little space. That's lots better than light airplanes that touch down at 40 to 80 miles per hour and then need enough space to brake to a stop. It's not hard to find a place to set a gyro down if you suddenly have to.

If a gyro does crash, there are certain ways in which it protects you. There's a vertical mast or rotor pylon above you that serves as a roll bar. The airframe is very strong for its size, and absorbs a crashing force by its "progressive collapse" effect. This is the way race car drivers at Indianapolis can survive collisions with a concrete wall at over 200 miles per hour. The structure of the vehicle yields with the force and cushions the blow, protecting the occupant from the impact of a sudden stop. In the same way, a gyro frame bends and parts break off in a crash. Each crunch of the machine means one less crunch for the pilot.

This gyro rolled over on a grassy surface, destroying the rotor blades, the propeller, and much of the airframe. The pilot was more embarrassed than hurt.

But there's one way in which the gyroplane is one of the least crashworthy aircraft in existence: It's those whirling rotor blades! All rotorcraft tend to thrash themselves to pieces when the rotor blades strike the ground or hit an obstruction. Fortunately, the parts tend to be thrown away from the pilot, who is near the center of the rotation. But this violent rotor blade thrashing tends to take parts of the machine with it, often breaking the mast and reducing other parts of the machine to scrap.

The odd thing is that as rotor blades have improved, their crashworthiness has become worse. The newer heavier, stiffer rotors provide more inertia and better lift while flying, but they don't break up easily while crashing. In the old days of lightweight wooden rotors it wasn't unusual for a tipped-over gyro to suffer no damage to the airframe or the rotor head. Machines were sometimes put back into service after nothing more than a meticulous inspection and a new set of rotors.

Today, a gyro that tips over slaps the ground with great force in its rotors. Something has to give, and it's usually more than the rotor blades.

Fortunately, the designers of gyros have already considered this problem and prepared their machines for a crash. Of course, that's not something you're likely to see in their advertising. You're not likely to be excited by an ad reading, "Our gyroplane crashes better than any other!" But it's something a good designer will be glad to talk to you about.

Every major gyro designer I've talked to has been willing to tell me in detail what happens to his machine in a crash. Good designers figure out in advance where the machine will yield, and they study any actual crashes of their machine to make sure their analysis is right. Somewhere in every good machine there's a "tear on the dotted line" point, hopefully placed where the pilot is not likely to get hurt.

Can I make my gyro stronger?

The answer is probably no. You probably can't make your gyro stronger if you're flying a machine that has good crashworthiness built into it. All the good designers I know fly their own machines, and all of them I've asked have experienced some sort of crash themselves. It's not that they are wild and crazy flyers. On the contrary, they are generally very cautious and conservative pilots. But they are their own test pilots, trying out new and unknown aircraft. They are also generally highly experienced, and every additional hour they've spent in the air is an additional chance for something to go wrong.

The designer of your gyro has his own neck on the line and has probably already done everything possible to make his aircraft as strong as possible in a crash.

I've seen unsuccessful attempts to improve on the strength of gyros. One man pounded hardwood dowels into the center of the square tubing of his gyro frame. He figured the added lumber would make the gyro so strong it couldn't possibly break in a crash. Maybe that wood-stuffed tubing won't break; maybe it will. But we do know that *something* will give in a crash—and on this machine we don't know where that will happen. He'll find out if his machine crashes, and he may not like what he learns! (Incidentally, there's nothing wrong with using a wooden core inside metal tubing, so long as the machine is engineered around this type of construction in the first place.)

Another example I've seen was a machine on which every bolt in the airframe was drilled and fitted with a cotter pin to assure that the nuts could not work loose. This was at least a waste of time and money, since aircraft lock nuts simply don't come off when they're in good condition and are properly tightened. It also probably produced a weaker structure due to the holes in the bolts.

If you have an idea of a way to make your gyro stronger, tell the designer. If it's a good idea he'll thank you for it and he will probably use it to help other people. It it's a bad idea he'll save you a lot of trouble by telling you so before you put a lot of work and money into it.

What causes gyro accidents?

Several studies have been made of gyroplane accidents, both fatal and non-fatal, with the same conclusion: The dominant cause is the pilot. If what bothers you is the thought of the machine coming apart in mid flight, you should know that this is less likely to happen than a wrong move by you. So maybe we ought to stop and think a bit about pilot error.

Pilot error

Actually, you can blame almost any type of accident on the pilot, since he's not only responsible for operating the machine, he's also responsible for the mechanical reliability of the aircraft. But there are ways a gyro pilot is more likely to be the cause of an accident that in other aircraft and other vehicles. Here are some of the reasons for pilot error:

Inexperience - All gyro pilots start with no experience with the special characteristics of this type of aircraft. While this is being learned, the pilot is at high risk of making a mistake. I can reliably predict that each pilot will make mistakes, but if he moves cautiously, they won't be big enough to hurt very much. Even with lots of dual-seat training, a pilot must advance slowly and cautiously in his flying.

To emphasize the point about advancing cautiously, several gyro instructors have told me that the students who are the most successful in their own single seat gyros are sometimes the most scared, over-cautious, slow-learning students during training. The instructors also have said that some of the bold, fast-learning students have gone home and crashed their own machines. There are no prizes for fast learning. Training away the pilot's inexperience is always a gradual process.

Unwillingness to fly - No matter how much they want to be gyro pilots, some individuals are not able to accept the different perspective that comes with flying. They may be in the air and flying, but they are afraid to get away from the ground and get away from the way you think in a ground vehicle. They may be unwilling to crab a little with the wind, reluctant to lean into a banked turn, unable to accept the idea of moving through the air, not on the ground. This can cause problems, like flaring for a landing while fifty feet up, losing airspeed in a downwind turn or flying too slowly, with the nose too high.

Bad training or too little training - In the past, a lot of the training of new gyro pilots was done by volunteer instructors who generously donated their time and effort to helping others learn. But despite their good intentions, not many of these people had any preparation to be instructors other than their own "hard knocks" experience. The volunteer instructors did their best, and sometimes their best wasn't enough to make their students competent. Today there's still some amateur training going on, and some of it is excellent. But now, with skilled professional instructors available, there's no need to settle for inadequate training.

It looks easy - You won't find this listed in FAA statistics on pilot error, but I personally think this is one of the greatest causes of pilot-induced accidents in gyros. Most gyros look like they'd be easy to fly. When you stand at the airfield and watch those choppers cavort around in the sky, you can easily overlook the hours and hours of practice, perserverence and perspiration that each of those pilots went through to get up there. Looking at all those differently-shaped machines, it's easy to conclude that anything will fly if it has rotor blades and an engine.

This has fooled people. There have been individuals with no flying experience who have climbed aboard gyros and attempted to fly them. They were fooled by the appearance of simplicity and ease (and perhaps by reassuring comments from other people who didn't know any better either). Not one of them succeeded. Not one.

Flying in a non-aircraft environment - Gyro pilots spend most of their time less than 200 feet above the ground. They often take off from fields that are not airports, where there are uncharted obstructions and unexpected hazards. Gyros usually don't fly in airspace regulated by the FAA, where flying patterns and procedures are well established. As a result, gyros are more likely than other aircraft to collide with power lines, with trees, with buildings or other unyielding objects. They are more likely to have a wheel drop into a gopher hole on takeoff or to roll over a patch of sticky mud.

Other pilot problems - There's no end to the ways pilots can get themselves into trouble in gyros. There's nothing to prevent a person who shouldn't be flying from trying it in a gyro, even though he may have impaired vision, poor muscle control, inadequate intelligence or some other deficiency that would prevent him from getting a pilot's license.

Here are five things that can help you minimize the likelihood of pilot error:

You need:

1) **Adequate knowledge** of gyros, of flying and of what it's all about,

2) **Adequate training** by a competent instructor,

3) **A slow, cautious approach** that puts no premium on getting things done quickly,

4) **Self examination** of how well suited you are to this special type of flying, and

5) **Defensive flying**, remembering that you're often in airspace where you're not expected.

Mechanical Problems

A less frequent cause of gyro accidents is having something break or stop working or come apart. Mechanical problems don't cause as many accidents as pilots do, but it's just as serious if it does happen. Here are some of the mechanical problems which have happened to gyroplanes.

Engine failure - This is Numero Uno on the top worry list of most pilots, whether they are in gyros or airplanes. Every good flight training program includes practicing very early for power-out landings. That's not just because your instructor wants to see you sweat. It's because engines are never perfect, and sometimes they quit. Fortunately, an engine failure is easy to handle in a gyro if the pilot is prepared. Nothing changes mechanically on a gyro in an engine-out condition. It's up to the pilot to fly it down to a safe landing.

Propeller failure - This is not the same as an engine failure. It's a much more serious matter. If any part of your propeller comes off, that part becomes a projectile with bullet-like speed which can do a lot of damage if it hits something. Worse yet, if the propeller is badly out of balance from the lost part, it can shake the engine loose from its mounts and even make the engine fall off. That would probably throw the machine so badly out of balance as to be uncontrollable. Obviously, it pays to take care of your prop!

Structural failure - With a rotorcraft it's somehow easy to imagine some sort of catastrophic breakup—the rotors come loose, the mast breaks off, the rudder tears loose, etc. In practice this hardly ever happens, due to the high quality of the gyroplanes that are available. In a gyro that's properly built and maintained, it's no more likely for the rotors to come off than it is for the wings of an airplane to tear loose. If it bothers you that your rotors are connected to your gyro with only one thick teeter bolt, take a look at the three or four little bolts that hold the wing on a light airplane. They are all built strong enough. If the gyro is competently designed, built and maintained, you can write off structural failure as a worry. If the gyro is not in good shape, don't worry about it. For goodness sake, *fix* it!

Broken controls - If any of the three controls goes haywire, you are in trouble. If the throttle were to break or stick or if the carburetor float were to break, you would have no control over the engine. That would mean the engine could shut down to an idle, stick at one throttle setting or go to full power. None of these is any fun. If a rudder cable were to break or if the rudder were to jam, you'd have trouble flying straight enough to fly or even land right. If the control stick were to jam or to get its linkages fouled, the machine might be hard to control or become totally uncontrollable. Problems with the controls are very rare occurrences, mainly because of the excellent problem-resistant control systems available today and the fact that people are concerned enough to take good care of their machines.

Unknown flight characteristics - A person who makes major modifications to his gyroplane or who hasn't learned enough about how these aircraft work may run into mechanical problems. For example, one pilot mounted a good-looking body enclosure on his gyroplane and proceeded to fly it

without problems. The body was actually designed for a
different machine. After lots of successful flights under
power, one day this pilot switched off the engine while in the
air. That's when he discovered an unknown flight characteristic: Without the propeller blast over the rudder, this modi-
fied machine was unstable. It turned every which way but
loose on the way down and crashed.

117

Bought a bad used machine - Buying used gyros is not for beginners. Unless you're an expert or have a true expert advising you, you have no sure way to know whether a second-hand gyro is airworthy or not. Most people who succeed at buying a used machine do a partial or complete tear-down of the machine, inspecting and replacing everything suspicious. In doing this, they've sometimes found substandard or unairworthy parts.

Pilot did not take responsibility - Here's a potential cause of mechanical failure you won't hear about elsewhere. But I'd nominate it as a major cause. In flying gyroplanes the total responsibility is carried by the pilot. You can't rely on someone else to warn you that something needs to be fixed or that you've done something risky to your machine. Unlike almost any other merchandise you can purchase, a gyro can't be guaranteed by anyone. Would you like a deal that reads, "Double your money back if it breaks and kills you?" There's no satisfactory recourse, no making good on a serious defect. It doesn't work to wait until your machine gives you a warning rattle or someone walks up to you and says, "Hey, did you know your rudder cable is frayed and about to break?"

Here are five things that can help you minimize the likelihood of mechanical problems:

1) **Learn all you can** about your gyro, how it works and how to maintain it;

2) **Fix everything** before you fly, never accepting any condition that's less than flight ready;

3) **Inspect your gyro** often and thoroughly, with a very critical eye;

4) **Don't invent anything** on your machine, at least not without having an expert check it; and

5) **Don't pass the buck;** Decide that nobody but you is in charge of your safety.

Am I Too Scared Now to Fly?

Good question. It's one you ought to stop and answer before you turn the page and read the next chapter.

If you're saying to yourself, "No way! I'm not going to have anything to do with gyroplanes," then you should be congratulated on making a very difficult choice, passing up something you probably wouldn't enjoy and opening the way to apply your energies somewhere else. If you're wavering, unsure whether or not this is for you, then keep reading and gathering information until you feel confident enough to make a decision. If you're saying, "Yes! I understand the risks and I know I can control them," then your carefully-considered choice is far more likely to lead you to success than a sudden impulse based on the sight of glittering rotor blades and the sound of a pulsating exhaust.

For many people, the risks involved in flying a gyroplane are appealing, not appalling. Some individuals are attracted by the freedom involved in this sport, which is not limited by the heavy-handed regulation that would be necessary to guarantee you near-perfect safety. These folks would rather take the responsibility for their own welfare, setting their own high standards and making sure they have protected their own safety and happiness.

There are people who have flown gyros for many years with success and safety. You can find people with gyro flying experience spanning 10 years, 20 years—even more than 30 years! Talk to these people and you'll find that they're not just lucky, they are people who have taken responsibility to see that gyro flying is not dangerous. For them it's less risky to be in the air in a gyroplane than to do ordinary things on the ground.

You've seen statistics at the beginning of this chapter indicating that sport gyroplane flying can be less risky than airplane flying. These are average figures. You can make your own chances even better than this.

Sure, there are things that can go wrong when you fly a gyro. And they can be pretty terrible. But when you do it right, the terrible things won't happen. Instead, what will happen are some very wonderful experiences. It's up to you!

Chapter 15

Pilot Report

This pilot report will attempt to answer the question of what it's like to fly one of these unique aircraft, the gyroplane. There won't be a lot of statistics, since flying a gyro is an experience, not an experiment. You don't fly a gyro to gather data. You do it to feel the sensations of flight in this most minimal of flying machines, with nothing around you but earth and sky, with no one to please but yourself.

So unimportant is data that some people fly these machines with absolutely no instruments aboard. They would rather judge their speed by the feel of the wind on their body and on the aircraft. They would rather reckon altitude by how much the landscape looks like a model train layout.

In some ways flying a gyroplane is what you expect it to be. In some ways it's not. You expect it to be exciting, windy and invigorating. You expect it to be different from a ride in an airplane. And it is. But in most ways gyro flight is not what you expect.

Surprises in store for you

The most important surprise when you fly a gyroplane is the discovery that "This is definitely not a toy!" Just looking at photos of gyros—and even just watching them fly—you can be reminded of those toy flying machines made out of balsa wood sticks and paper—or the kind that come in a G. I. Joe box with a figure of Cobra Commander sitting on a little plastic seat.

In fact, most gyros really are made out of sticks and bolts and a little motor, sort of like an Erector set. But when you fly one, you discover that these are massively strong

Does this look scary? You're in for a big surprise!

sticks and bolts, and that the little motor puts out more power than the engine in a lot of 2,000-pound cars—definitely not a toy!

Those rotor blades are spinning hundreds of miles per hour, and that machine is racing through the air at speeds that could get you arrested on a highway—definitely not a toy!

The gyroplane easily lifts you hundreds or thousands of feet into the sky and responds powerfully to your slightest pressure on the control stick. You *know* this is not a toy!

Another surprise you'll find is the sense of security you have aboard a gyro in flight. To many ground observers a gyro looks like a wild collection of violent random movement, with the rotor blades whirling one way, the engine and propeller spinning another way, and all the parts seemingly on the verge of flying off in different directions. But from the pilot's seat, near the center of all this movement, you discover that all this action is perfectly coordinated and in harmony.

You're in for another surprise if rotor blades look to you like a pretty flimsy set of wings. When rotors spin they seem to disappear, leaving you literally with no visible means of support. People sometimes notice this and feel uncomfortable when riding a helicopter, but a gyro pilot has no such insecurity. With the control stick in your hand, you feel the presence and the power of the rotor blades. From where you sit, you can see that those rotor blades are creating a huge disc with more square feet than some living rooms.

You'll probably be surprised by the sound of the rotor blades. While people on the ground may call it a "chopper," you'll only hear the steady swish of those blades from the pilot's seat. They sound like they want to keep spinning forever (which happens to be quite true, since nothing will stop those blades or even make them vary more than ten percent in speed until your flight is over).

Another thing that will probably surprise you is the air blast you feel at flying speed. On open-frame machines there's nothing to absorb the force of this blast, since your body is the first object to break the 50-mile-per-hour wind. It's stronger than the feel of 50 miles per hour on a motorcycle, where the body of the bike knocks a lot of air away from you.

To get an idea of what it's like, watch closely when a gyro flies near you. At cruising speed the pilot's pants legs will be rippling furiously, giving you a clue to the strength of the blast. Though the force is strong, it's not unpleasant, and some people find it especially refreshing on hot summer days. But it is surprising.

Have you seen a gyro floating by, seeming to glide through the air effortlessly with the engine purring and the rotor blades singing perfectly smoothly? Unfortunately, you won't experience that impression of smoothness in the air. There are lots of little bumps up there—gusts, air pockets, updrafts, downdrafts and the like. And except for those rare days when the air is perfectly calm and glassy smooth, you feel those bumps as you fly, just as you do in any aircraft.

Actually, considering that your aircraft weighs about one third as much as a Cessna 152, it's surprising that the gyro gives you a smoother ride. That's because rotor blades have a way of cutting through the imperfections in the air like a wing with a heavy wing loading. In fact, your gyro ought to ride more like a Bonanza than a Cessna 152, if you figure wing loading the way airplane engineers do (gross weight divided by the wing area). In rotorcraft this is called blade loading (gross weight divided by rotor blade area). Your gyroplane blade loading is over 40 pounds per square foot, while the Cessna 152 wing loading is only a little over 9 pounds per square foot. (A Bonanza wing loading is still less than half that of a gyroplane at about 16 pounds per square foot.)

Besides feeling the bumps in the air, you'll also feel vibration from the engine and the rotor blades. Even the best balanced and trimmed rotor blades shake a little as they spin, producing a vibration that you feel in the stick. Helicopters have fancy damping devices to keep passengers from noticing this rotor shake, but in a gyroplane you and the rotor are bolted solidly together. When you spin a twenty-odd-foot-long rotor blade at around 400 revolutions per minute, even microscopic variations in weight or shape produce vibrations you can feel. In a good machine the rotor shake is minimal, but it's always there.

Flight characteristics

You can sum up gyro flight characteristics in a sentence: It feels like it wants to play! It can make a 45-degree banked turn as easily as it can fly straight and level. It can climb or descend rapidly, and can change from one to the other with no ceremony or hesitation. Punch the rudder and you can find yourself flying nearly sideways (Those rotors don't care which way the gyro is going!) Fly tight circles around a spot on the ground like a bird. Chase a bird if you like. It all feels comfortable in a gyro!

You might think this means the gyroplane is no fun to fly straight and level. No, indeed! This is one of the most enjoyable aircraft you could fly that way, because it feels so comfortable and stable. A properly-trimmed gyroplane will fly with your hands off the controls (It will even fly with hands and feet off!). On a cross-country trip, a gyro pilot can devote nearly all of his attention to sight-seeing (or watching for an emergency landing spot, since no engine is perfectly reliable), with very little effort spent handling the aircraft.

This machine responds quickly and powerfully to any control input. But how it responds depends a lot on airspeed. At cruising speed the stick (the "cyclic control" in rotorcraft terms) is sensitive and powerful, since it makes the entire rotary wing tilt, not just an aileron. Only slight pressure is needed to get a response in pitch (nose up/nose down) and roll (tilt left/tilt right). The rudder is also quite sensitive at cruising speed, since the airframe offers very little resistance to yaw movement (nose left/nose right).

At low speeds of 20 to 30 miles per hour, such as on takeoff, the controls aren't so sensitive. If you watch a gyro on takeoff or landing, you may be able to see the pilot move the stick, since his inputs are fairly large. The rudder remains powerful and effective at low speeds because it sits right smack in the middle of the propeller blast.

This change in control response, from loose at low speeds to tight at fast speeds, works to the pilot's advantage. On takeoff and landing, when the pilot needs a forgiving machine with a lot of rudder power to stay lined up, the gyro acts just the way you'd want. Then, at cruise, when you don't want to work so hard at controlling the machine, it tightens up and allows you to handle it seemingly with thought waves.

Coordinating the rudder with the stick on turns is no problem. In fact, except for takeoffs and landings, when careful rudder control is needed, you can forget the rudder altogether and fly with just the stick. The machine will "weathercock," automatically pointing in the direction you lead it. Airplanes require use of the rudder on turns to overcome "adverse yaw," but gyroplanes have no ailerons and no adverse yaw.

The playful, responsive nature of the gyroplane is sometimes a temptation to push the machine beyond the pilot's abilities. A pilot's skill is magnified by the responsiveness of this fun little bird, making it seem that you can do more than you really should. You have to hold back with this little mistress, think of Mom and the kids, and always stay within your own proficiency.

How quick is the gyro?

Without a doubt, the gyroplane is a quickly responding aircraft. But just how quick is it?

Different gyroplanes respond at different rates. During your first flights in a deal-seat trainer, you're riding in a fairly slowly-responding machine. It's heavy, the rotor blades are long, and you're probably not flying a high speeds.

When you fly solo, things are different. Even in the big two-seat gyro, your reduced weight allows it to respond faster. And in a smaller single-seat machine it gets quicker still. But surprisingly, even a single-seat gyroplane won't be as quick to respond as a small homebuilt airplane.

It seems odd that a little 500-pound gyroplane can actually be no more touchy than an airplane weighing lots more. There are a couple of reasons for this: First, the spinning rotors of a gyroplane create quite a gyroscopic effect, which resists being tilted the way a toy gyroscope does. Second, while the control stick of an airplane instantly changes the lift generated by each wing, the control stick of a gyroplane doesn't immediately move the rotors. What it does is tilt the rotor head, and the rotors follow. There's a delayed response due to a phenomenon called phase lag. This means that the blades must spin as much as 90º around the mast before they move to a new position.

A gyroplane handles a lot like an airplane. You fly a gyro with just stick, rudder and throttle. The controls aren't complicated like those of a helicopter, where it seems as though you have to rub your stomach, pat your head and wiggle your ears at the same time.

But somehow a gyroplane doesn't *feel* like an airplane. That's because an airplane is basically a collection of fins that work together in cutting the air, while a gyroplane is a frame hanging from a canopy.

The two kinds of aircraft are completely different in the way they are stabilized aerodynamically. That's why you can't fly a gyroplane without special training, no matter what you've flown before.

When you're up there in a gyro, you *know* you're not in an airplane—and once you've got the hang of it, you're *glad!*

Let's go flying!

To make this a complete report, why not take a flight in a gyroplane? Let's see what it feels like and observe what happens. And to make it as exciting as possible, we'll make our flight in one of the open-frame machines.

Let's start by imagining that you are some kind of Superman, capable of running two or three times as fast as the quickest Olympian, without the slightest physical effort. But you do this in a special way: sitting down in a lawn chair. Put your lawn chair down at the end of a runway and take in your hands the special stick that works like a magic wand, giving you immense power to direct the chair with your finger tips.

With a nudge of that magic wand, you begin to run forward effortlessly, quickly passing into superhuman speed. Another move of the wand allows you to run right off the ground and into the air. Now you're sitting in a chair in the air with the ground falling away, seeming to float slower and slower as you climb.

There's no effort to it, as you begin to feel more a part of the sky than the earth. As the ground recedes, it continues to slow down, reaching the point where you seem to be just creeping along through the air. Seemingly in contradiction of this slow-motion appearance, the air keeps blasting at you, pushing on your helmeted head and rippling your clothes. If you had any loose objects in your pockets, they've been blown overboard by now.

You look around yourself, expecting to see walls and windows that would enclose you in a car or an airplane. Your senses are shocked as you look at your feet in front of you. Between your feet is the keel of your machine. Around you is *nothing!* Nothing but green and brown fields, tiny houses and toy barns, all looking ever more perfect and clean as they shrink smaller and smaller.

Whatever flying machine there is, is above you and behind you where you don't see it. It makes you feel as though you yourself are the flying object, not just a passenger in a vehicle. You have the sensation of being in a seat floating through the sky. It can tilt and swoop and move wherever you tell it to go, just by moving the magic wand in your hands.

By now you may be several hundred feet up. You're actually as high as a skyscraper, but you don't feel it. You're comfortable while you're looking at the horizon, keeping your flying chair pointed where you want it to go. But when you glance down at the ground directly below you, you discover it's a long way down there! "Holy acrophobia!" you say, and resolve not to look straight down again.

Now that you've gained altitude, you notice a strange phenomenon: That seat that was plenty wide on the ground has gotten a lot narrower! And the higher you go, the narrower it gets. Nobody knows for sure why this happens, but any gyro pilot can attest to it.

As you look over your feet at your intended landing strip, you may wonder how those big feet—and this even bigger machine behind you—will fit into that tiny-looking space on the ground. But since the world seems to be moving by so slowly from way up here, you figure maybe your special Superman powers will help you get down.

You reduce throttle and descend. Not surprisingly, the ground appears to pick up speed as you get closer to it. Soon you're zipping along at 50 miles per hour a few feet above the strip. You gradually pull back the throttle and flare, more and more steeply until the nose comes up high and it feels like you just leaped seat-first into a big marshmallow cushion.

Those magnificent rotor blades speed up and bite into the air like a big drag parachute, gently setting you down so softly that you almost feel as though you could put your feet down and make a stand-up landing. Inside your mind, this landing reminds you of childhood dreams of jumping into a giant glob of Jell-O or a big pile of cotton candy or . . . something soft as air!

And that's where a gyroplane flight ends: in your mind. Echoing around inside your head are all kinds of pleasant sensations. Right after landing there's not a speck of data in your brain. Whatever you used to worry about is completely gone from your head. You are overwhelmed by the experience.

A little while after the flight you begin to recall the data and can describe it to your friends in appropriately intellectual terms. And little by little, you start to remember those ordinary problems you used to worry about. Your Superman image fades away and you discover that you are still an ordinary Earth person, after all.

...until your next gyro flight!

Between your feet is the keel of the machine. Around that is nothing but green and brown fields, tiny houses and toy barns.

Chapter 16

I Can Fly! What's Next?

What do you do after you've made your first solo flights around the pattern? Obviously there's the second, third and umpteenth trip around the pattern to make. But what then?

To be a consistent and safe gyroplane pilot, you need to have enough experience to be able to handle anything that may happen. This includes practice in emergency landings. Regardless of how proficient you are, you need to continue to practice these. Simply make a high approach over your landing spot, reduce power to an idle, and land. The airspeed should remain near cruise speed during your power off approach, all the way until you level out about one or two feet above the ground.

While practicing, add power if needed to get down comfortably, but work toward bringing the aircraft to a landing with the engine idling all the way. (There's no need to actually shut off the engine since the machine handles about the same with the engine idling, and this gives you the option of applying power if things get tight.)

You should be able to predict your landing spot as soon as you reduce power and to maneuver to make your

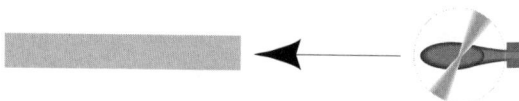

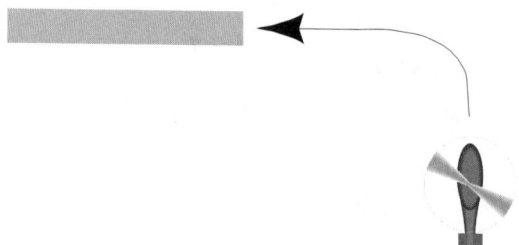

Noted pilot, designer and manufacturer Ken Brock flew his gyroplane from coast to coast and also reached an altitude of 13,500 feet.

prediction come true. You should always know which way the wind is blowing and always try to land into the wind.

After you've made a few power-off approaches landing straight ahead, try landing "around the corner." Reduce power while you still have a turn to make before you are lined up to land. Then if it appears you are going to land short, you can cut the corner and make your spot. If you're going past your spot you can lengthen the corner to make it. Make sure you complete the turn and straighten out before touching down, since you can't land in a turn. You'll be glad to have this experience if and when the day comes when you actually need it.

There's lots more to learn about flying a gyroplane: slow flight, vertical descents, high-performance takeoffs and even cross-country trips. They all await you after you have mastered normal flight.

There are so many things you can do with a gyro, like attending fly-ins, making fly-bys for ground-to-air photography, learning new maneuvers, up-grading your equipment and . . . There's enough for a whole lifetime of gyroplane flying.

The gyroplane is an amazing machine. After flying one, it's hard to understand what happened to the tremendous popularity of these aircraft during the 1930's and early 1940's.

Gyroplanes are responsive, sporty and exciting. People who fly everything from giant airliners to helicopters have said that gyroplanes are the most enjoyable aircraft to fly.

But fun as it is, the gyroplane is slower on a cross-country trip than an airplane. The gyroplane won't land in as short a space as a helicopter. It won't carry a payload very efficiently. No one has yet found a good commercial use for the gyroplane.

The one and only gyroplane claim to superiority is the fun of flying it, and the remarkable ease of building and piloting the gyroplanes that are available today. Until recent years that was not enough to overcome the gyroplane's lack of commercial success. But in the last few decades the era of sport flying has arrived, the era of aircraft designed and built solely for the joy of flying.

Now people venture into the air in ultralight airplanes, in hang gliders, in balloons and even in tail-first foam plastic airplanes. Now nobody cares if the new sport aircraft won't haul mail or passengers for a profit.

What counts now is simply how well the flying machine fits the flyer. In this era, the gyroplane has at last found its chance for success. And you, my friend—you and your gyroplane—you are part of it all!

Index

Index